Marc A_____
Prize C_____ e science-humour
magazine *Annals of Improbable Research*, runs the
popular website <u>www.improbable.com</u>, and writes
a weekly column for the *Guardian*. Ten new Ig™
Nobel Prizes are awarded every year. These are
physically presented to the astonished winners by
genuine Nobel Laureates in a lavish ceremony at
Harvard University.

By Marc Abrahams

Ig™ Nobel Prizes: The Annals
of Improbable Research

Ig™ Nobel Prizes 2: Why Chickens
Prefer Beautiful Humans

IG NOBEL PRIZES 2

Why Chickens Prefer Beautiful Humans

Marc Abrahams

ORION

An Orion paperback

First published in Great Britain in 2004
by Orion
This paperback edition published in 2005
by Orion Books Ltd,
Orion House, 5 Upper St Martin's Lane,
London, WC2H 9EA

1 3 5 7 9 10 8 6 4 2

A CIP catalogue record for this book is
available from the British Library.

ISBN 0 75286 461 0

Typeset by Butler and Tanner Ltd, Frome and London

Printed and bound in Great Britain by
Clays Ltd, St Ives plc

www.orionbooks.co.uk

This book is for my parents

Contents

Acknowledgements

Special Ig thanks to my wife, Robin Abrahams, and to Sid Abrahams, Margot Button, Sip Siperstein, Don Kater, Stanley Eigen, Jackie Baum, Joe Wrinn, Gary Dryfoos, the Harvard-Radcliffe Society of Physics Students, the Harvard-Radcliffe Science Fiction Society and the Harvard Computer Society, the Sanders Theatre production staff, the Harvard Box Office staff, the Harvard Extension School and the MIT Press Bookstore.

Special publishing thanks to Regula Noetzli and Ian Marshall.

The Ig Nobel Prize Ceremony and this book would not have been possible without a lot of help from a large number of people. Among them:

Sondra Allen, the American Association for the Advancement of Science, Alan Asadorian and Dorian Photo Lab, referee John Barrett, Charles Bergquist, Doug Berman, John Bradley, Silvery Jim Bredt, the British Association for the Advancement of Science, Alan Brody, Jeff Bryant, Nick Carstoiu, Jon Chase, Keith Clark, Frank Cunningham, Investigator T. Divens, the Dresden Dolls, Bob Dushman, Kate Eppers, Steve Farrar, Dave Feldman, Len Finegold, Ira Flatow, Stefanie Friedhoff, Jerry Friedman, Shoichi Fukayama, Shelly Glashow, Mary Hanson, Jeff Hermes, Dudley Herschbach, David Holzman, Karen Hopkin, Jo Rita Jordan, Roger Kautz, Gerard Kelly, David Kessler, Wolfgang Ketterle, Hoppin' Harpaul Kohli, Alex Kohn, Leslie Lawrence, Jerry and Maggie Lettvin, Barbara Lewis, Tom Lehrer, Harry Lipkin, Colonel Bill Lipscomb, Julia Lunetta, Counter-clockwise Mahoney, Lois Malone, prominent New York attorney William J. Maloney, Mary Chung's Restaurant, Micheline Mathews-Roth, Les Frères Michel, Lisa Mullins, the Museum of Bad Art, Steve Nadis, Greg Neil, the Flying Petscheks, Harriet Provine, Genevieve Reynolds, Rich Roberts, Nailah Robinson, Bob Rose, Katrina, Natasha, Sylvia, Isabelle and Daniel Rosenberg, Louise Sacco, Rob Sanders, Annette Smith, Miles Smith, Kris Snibbe, Naomi Stephen, Mike Strauss, Alan Symonds, Doug Tanger, Al Teich, *The Times Higher Education Supplement*, Peaco Todd, Tom Ulrich, Paula Wallace, Verena Wieloch, Bob Wilson, Eric Workman, Howard Zaharoff, and thank you, thank you, thank you to Martin Gardner.

And, of course, thank you – on behalf of all of us – to the Ig Nobel Prize winners, and to everyone who has sent in nominations.

Introduction

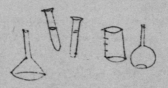

Implausible, Improbable and True

That any one of these things happened seems implausible. That all of them happened seems improbable. But they did:

- A man in India started a political movement because he was legally dead
- Tests in Stockholm, Sweden, showed that chickens prefer beautiful humans
- It became possible to rent an entire country (Liechtenstein) for corporate conventions, weddings and bar mitzvahs
- An Englishman compared scrotal asymmetry in man and ancient sculpture
- Scandinavian volunteers assessed the impact of wet underwear on thermal comfort in the cold
- A Swiss-Japanese-Czech team noted the effects of chewing-gun flavour on brain waves
- Three physicians agreed on guidelines for dealing with the zipper-entrapped penis
- Two brothers figured out how to insinuate pungent scratch-and-sniffable perfume into magazines
- A Japanese device translated dog barks into human language
- An American engineer calculated that Mikhail Gorbachev is the Antichrist, and
- A Dutch ornithologist observed homosexual necrophilia in the mallard duck.

Each of these achievements deserves acknowledgement of some kind, from someone. Each got it, duly and with ceremony, in the form of an Ig Nobel Prize.

This is the logo of both the Ig Nobel Prize Ceremony and the *Annals of Improbable Research*, the magazine that documents all things Ig Nobel.

What does it take to win an Ig Nobel Prize? There is but a single criterion: do something that first makes people *laugh*, and then makes them *think*. (*What* people think is entirely up to them.) Until recently, we used more elaborate language to describe this. I am mildly embarrassed that it took us about twelve years to boil it down to a simple phrase that anyone can understand.

I founded the Ig Nobel Prize Ceremony, with help from many other people, and serve as chairman of the Ig Nobel Board of Governors and as master of ceremonies when the Prizes are handed out. I am also the editor of a science humour magazine called the *Annals of Improbable Research*, also known as *AIR*. *AIR* is the publication that officially documents each year's crop of Ig Nobel winners.

This book tells the story of a good, round helping of Ig Nobel winners – what they did and why they did it. In a previous book, *Ig Nobel Prizes*, I did something similar for some of the other winners.

Each of these people can well and truly be called a piece of work. Of most of them, I say that admiringly. Each probably deserves to have an entire book written about him or her. Some already have had books written about them. One, Troy Hurtubise of anti-grizzly-bear suit-of-armour fame, has been the star of a documentary film, and Troy's story is so very unlike any other that I could not resist including him in both books. Thus the chapter here called 'The Further Adventures of Troy'.

A complete list of the previous winners is in the appendix at the rear of this book. Perhaps you will help us add to the list, either by nominating a future winner or, if fate insists, by becoming one yourself.

The Ig Nobel Prize

An Ig Nobel-winning achievement can be good or bad, important or trivial, valuable or worthless. Or all of those. It just has to make people laugh, and then make them think.

We award ten Prizes every year. The First Annual Ig Nobel Prize Ceremony was held in 1991. The Thirteenth First Annual Ig Nobel Prize Ceremony was held in 2003. As of this writing, 130 Prizes have been awarded.

It is not easy to win an Ig Nobel Prize. Every year we receive more than 5000 nominations. To select just ten is an excruciating, if unimportant, task. Whatever doesn't win a Prize one year goes back into the pool for consideration the next. So the odds against winning are stout.

It would be all too easy to select ten winners every year from the United Kingdom alone. But that would not be fair to the citizens of other nations. Of course it would also be easy to select ten winners from Japan, or from the United States, or from a number of other countries. Fortunately, the competition is fierce, whether the competitors know it (or even know they are competitors) or not.

Somewhere between 10 and 20 per cent of the nominations come from people who are nominating themselves. But so far only one Prize has been awarded to a self-nominee. That went to Anders Barheim and Hogne Sandvik of the University of Bergen, Norway, who won the 1996 Ig Nobel Biology Prize for their tasty and tasteful report, 'Effect of Ale, Garlic and Soured Cream on the Appetite of Leeches', which was published in the *British Medical Journal*, volume 309, 24–31 December 1994, on p. 1689.

If you are selected to win an Ig Nobel Prize, you will usually be given the opportunity to quietly decline the honour. Only a handful of individuals have turned it down; in almost every case someone – typically a boss – was already out to get the Ig nominee, who didn't want to give that someone yet another excuse to be a bully.

There have been many years when we failed to get in touch with the Economics Prize winner, because he or she had a previous five-to-fifteen-year engagement. So it goes.

The Ceremony

Each winner receives an invitation to take part in the Ig Nobel Prize Ceremony, at Harvard University. They have to travel at their own expense, but most find it worthwhile. The Ceremony is held in Sanders Theatre, Harvard's oldest, largest, and by far most dignified meeting place. On Ig Nobel night the place is always packed to the rafters with a sell-out crowd of 1200, many of whom spend the entire evening wafting paper airplanes at the stage. The people onstage wing them right back.

The heart of the Ceremony comes when each of the ten new winners is announced. A winner steps through the Sacred Curtain at centre stage, whereupon a Nobel Laureate (yes, a genuine Nobel Laureate) shakes his or her hand and presents the Ig Nobel Prize. All parties are visibly delighted and impressed, sometimes in mildly euphoric shock, evidenced by giggling and rictus.

The Prize itself is hand crafted, of a new, different design every year, and always made of exceedingly cheap materials. Each winner also gets a certificate attesting to the fact that he or she won an Ig. The certificate is signed by several Nobel Laureates.

We like to keep the Ceremony moving along. Each winner is allotted a maximum of 60 seconds to deliver an acceptance speech. To ensure that nobody drones on, a cute eight-year-old girl sits on one side of the stage, monitoring events. We call her 'Miss Sweetie Poo', and introduce her at the start of the Ceremony, explaining that whenever Miss Sweetie Poo feels that someone has talked long enough, she will let that person know. Miss Sweetie Poo gets her message across simply and directly. She walks across that stage, stands right next to the person who's yacking away at the microphone, and says, 'Please stop. I'm bored. Please stop. I'm bored. Please stop. I'm bored. Please stop. I'm bored...' No-one outlasts Miss Sweetie Poo. Few dare try.

The Ig Nobel Prize Ceremony includes a variety of other delights, too. There's the Win-a-Date-With-a-Nobel-Laureate Contest, in which one lucky audience member wins a date with a Nobel Laureate.

Eight-year-old Miss Sweetie Poo informs Ig Nobel Biology Prize winner Charles Paxton that his acceptance speech has gone on long enough. Paxton and three colleagues were honoured for their report 'Courtship Behaviour of Ostriches Towards Humans Under Farming Conditions in Britain'. (Eric Workman/*Annals of Improbable Research*)

There are the Nano-lectures, in which several of the world's great thinkers explain their favourite subjects, first in a fast-talking 24 seconds, and then as clearly and understandably as possible in 7 words. The 24-second time limit is strictly enforced by Mr John Barrett, a professional football referee. Mr Barrett is reinforced by Mr William J. Maloney, a prominent New York attorney, equipped with a tiny tin trumpet. Mr Maloney monitors all utterances to ensure that nothing offensive reaches the ears of the audience.

And each year we write a mini-opera about some topic in science. The mini-opera is sung by full-sized

A lucky audience member claims her prize (chemist Dudley Herschbach) in the Win-a-Date-With-a-Ig Nobel Laureate contest. (Jon Chase/Harvard News Office)

professional opera singers accompanied by Nobel Laureates.

The Ceremony has many other parts, too – all moving, thanks to the work of Miss Sweetie Poo and her colleagues, with pleasing rapidity.

The entire thing is televised live on the Internet. On our website, www.improbable.com, video of several past Ceremonies has been posted.

Two days after the Ceremony, the winners, by then relaxed and somewhat in possession of their normal faculties, give an afternoon of free public lectures. This happens at the Massachusetts Institute of Technology, two miles down the road from Harvard. The audience always becomes very involved in questioning the winners, and the winners get very excited proposing various ways in which the bunch of them might collaborate. This can be both hilarious and, at least in theory, frightening (glance over the contents of this book and you will quickly see why).

The Ig Nobel Prize Ceremony occurs on, or within a week of, the first Thursday evening in October.

Part of the audience of 1200 at the Ig Nobel Prize Ceremony. The figure at right is a giant stone statue – one of a pair that loom from the sides of the stage in Sanders Theatre. (Eric Workman/*Annals of Improbable Research*)

The Ig Nobel Tour of the UK and Ireland

Every March, several Ig Nobel Prize winners and I do a barnstorming tour of cities in the United Kingdom and Ireland. The British Association for the Advancement of Science organizes this as one of the main features of their National Science Week. The first two tours were generously funded and co-organized by *The Times Higher Education Supplement*.

It is an oddity of the UK's National Science Week that

Two days after The Ig Nobel Ceremony at Harvard University, the new winners give an afternoon of free public lectures. The audience pays rapt attention. This action photo was taken by Ig Nobel Prize winner C.W. Moeliker at the 2003 Ig Informal Lectures.

Pek Van Andel, leader of the Dutch team that won the 2000 Medicine Prize, explains how and why they took MRI images of a couple's sexual organs while those sexual organs were in use. This photo was taken by Ig Nobel Prize winner C.W. Moeliker at the London event of the 2004 Ig Nobel Tour of the UK and Ireland.

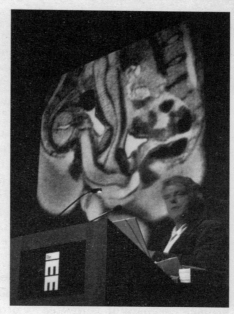

it stretches over nearly two weeks and is not entirely confined to the UK. The 2004 tour included shows in nine cities – Oxford, Nottingham, Belfast, Dublin, Glasgow, Exeter, Manchester, London and Warrington.

C.W. Moeliker's stuffed mallard duck, in a Dublin pub during the 2004 Ig Nobel Tour of the UK and Ireland. Moeliker won an Ig Nobel Biology Prize for producing the scientific documentation of homosexual necrophilia in the mallard duck. (C.W. Moeliker)

In a typical show, each Ig winner gave a five-minute talk and took *lots* of questions. When the time limit is strictly enforced, a five-minute talk can be wonderfully enter-taining and enlightening, especially if the topic is 'Scrotal Asymmetry in Man and in Ancient Sculpture', or 'Courtship Behaviour of Ostriches Towards Humans Under Farming Conditions in Britain', or 'The First MRI Images of a Couple's Sexual Organs While Those Sexual Organs Were in Use', or 'Evidence That London Taxi Drivers Have Brains That Are More Highly Developed Than Those of Their Fellow Citizens', or 'The Possible Pain Experienced During Execution by Different Methods', or 'Homosexual Necrophilia in the Mallard Duck', 'Using Magnets to Levitate a Frog', or 'How to Rent Liechtenstein'.

Most shows on the tour also included a performance of one of the Ig Nobel mini-operas and, in cities where British self-restraint permitted, a profusion of paper planes from the audience.

Similar Ig Nobel Tours are starting up in Australia and other countries.

Controversy

The Igs have occasionally been the subject of controversy.

Sometimes this involved seekers, disappointed individuals who mounted elaborate, long-running campaigns to secure a Prize for themselves, or in some cases, for a rival.

Sometimes it was dissenters, worshipful admirers of certain Ig Nobel winners, who protested angrily that their man deserved something higher, grander than an Ig Nobel Prize.

In 1995, it was the United Kingdom's Chief Scientific Adviser. Sir Robert May sent two letters to the Ig Nobel Board of Governors demanding that we stop giving Ig Nobel Prizes to British scientists – even if (as was the case) those scientists wanted to receive the prizes. Dr May's letters were triggered, somehow, by the awarding of the 1995 Ig Nobel Physics Prize to a team of British scientists who had explored the physio-chemical process of how, exactly, breakfast-cereal flakes grow soggy. The scientists had been pleased to receive an Ig, although they had not expected that the one person *not* to be pleased would be their government's highest science official. A tiny tempest ensued. Newspapers, magazines, and science journals ran amused editorials saying that the Ig Nobel Prizes were a good thing, and that defensive bureaucrats were perhaps not such a good thing. *L'affaire* May, like the various little campaigns for and against particular Ig Nobel selections, was that lovely sort of controversy in which no-one emerges the least bit harmed.

In 2002, Sir David King, Robert May's successor in the post of Chief Scientific Adviser, came to the Ig Nobel Prize Ceremony, helping hand out the prizes, and also singing, after a fashion, in that year's mini-opera. The following spring he took part in the Ig Nobel Tour of the UK.

How to Nominate Someone

REQUIREMENT FOR WINNING A PRIZE
Ig Nobel Prizes are given for achievements that first make people *laugh*, and then make them *think*.

WHO IS AUTHORIZED TO SEND IN NOMINATIONS
Anyone, including you. Especially you.

WHO IS ELIGIBLE TO WIN
Any individual or group, anywhere.

WHO IS NOT ELIGIBLE TO WIN
People who are fictional or whose existence – and achievement – cannot be verified.

CATEGORIES
Each Prize is given in a particular category. Some categories recur every year – Biology, Medicine, Physics, Peace, Economics. Others (Safety Engineering, Environmental Protection) are created to fit the peculiar nature of a particular achievement.

HOW TO SEND IN A NOMINATION
Gather information that explains *who* the nominee is,

and *what* the nominee has accomplished. Please include enough information that the judges can get an immediate, clear appreciation of why the candidate deserves an Ig Nobel Prize. Also indicate specifically where the judges can find further information if they need it, including (if you know it) how we can get in touch with the nominee. Mail or e-mail the nomination to:

Ig Nobel Nominations
c/o *Annals of Improbable Research*
PO Box 380853
Cambridge MA 02238 USA
air@improbable.com

If you mail the material and would like a response, please include an e-mail address or an adequately stamped, self-addressed envelope. If you wish anonymity you can have it. The Ig Nobel Board of Governors typically loses or discards most of its records, anyhow.

You can find further information at the *Annals of Improbable Research* website (www.improbable.com). There you can also sign up for the free monthly e-mail newsletter.

Sentiment

As we say at the end of each year's ceremony: 'If you didn't win an Ig Nobel Prize – and especially if you did – better luck next year!'

1. Larger Than Life

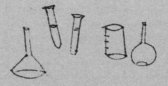

Some people really are larger than life. For one man, a single lifetime proved to be but the first in a series. Another man became widely famous, though many thought him merely mythical. A third man heroically protects himself against grizzly bears and other superhuman forces.

Behold them and their works.

Knight of the Living Dead

The Birth of Murphy's Law

The Further Adventures of Troy

Knight of the Living Dead

The most famous member of Mritak Sangh is undoubtedly Lal Bihari, the founder. When he applied for a bank loan in 1975, he discovered he had been declared dead by his uncle, who had bribed to get a death certificate and then bribed again at the land registry office to get the land transferred. How do you prove you are actually alive, especially if the case is stuck in court for 25 years? Lal Bihari tried publicity – trying to get arrested, standing for Parliament, attracting contempt charges by courts, writing pamphlets, organising his funeral, demanding a widow's pension for his wife, adding the word Mritak ('dead') to his name. All of these would have publicly acknowledged that he existed. It took 19 years, from 1975 to 1994, for him to prove he was alive, thanks to help from the District Magistrate. Having been reincarnated, Lal Bihari became the champion of other living dead.

— Bibek Debroy, writing in the Indian newspaper the *Financial Express*, 7 January, 2003

A grainy photograph of a public rally held by Lal Bihari (seen here wearing a black vest, but no skeleton costume) and fellow members of the Association of Dead People. (Photo courtesy of Lal Bihari)

The Official Citation

The Ig Nobel Peace Prize was awarded to:

Lal Bihari, of Uttar Pradesh, India, for a triple
accomplishment:
First, for leading an active life even though he has been
declared legally dead;
Second, for waging a lively posthumous campaign against
bureaucratic inertia and greedy relatives; and
Third, for creating the Association of Dead People.

IGNOBEL PRIZES 2   Lal Bihari's life and death have
been chronicled in many
newspapers and magazines in India and
elsewhere. Here are three of the articles:
High Court Rescues 'Living Dead' of Azamgarh,
Indian Express, 25 July, 1999.
Back to Life in India, Without Reincarnation, *New
York Times*, 24 October, 2000.
Mritak Sangh and Lal Bihari, the *Financial
Express*, 7 January, 2003.
The Indian film maker Satish Kaushik is filming a
movie about Lal Bihari.

Several years after his death, Lal Bihari came up with
the idea that publicity might do him some good. This
proved to be correct.

The members of the Association of Dead People are
biologically quite alive. But bureaucratically they are
kaput. Each had succumbed to what might be called
'murder by paperwork'. Their demise by certificate is
what qualified them for membership of the Association.

In each case, some greedy relative or friend had stood to inherit property, and had bribed a compliant government official to certify the victim's death. Though this was illegal, it was easy to do.

When Lal Bihari was a young man, his uncle had him declared legally dead. Suddenly landless, and legally lifeless, Lal Bihari tried unsuccessfully to regain what had been taken from him. Over the next several years, he discovered that many other residents of the province of Uttar Pradesh shared his peculiar state of decease. Surmising that unorthodox methods might be the best hope for all of them, Lal Bihari founded his organization. Its name is spectacularly unspectacular: 'Mritak Sangh', which translates in English to 'Association of Dead People'.

Bary Bearak, a *New York Times* reporter, visited Lal Bihari in the year 2000, and published a report saying:

His legal resurrection was accomplished in a mere 19 years, and in the process, Mr Bihari, a poorly educated merchant, found his mission in life: championing the cause of the similarly expunged. In July, a High Court judge became aghast after learning that there were dozens – and perhaps hundreds – of such cases of bogus mortality. He ordered the government of Uttar Pradesh to publish ads, seeking out the living dead, and then to revive them in the state's public records. The National Human Rights Commission has also convened hearings on the matter. 'As the bureaucrats once feared the devil, they now fear the Association of Dead People,' said Mr Bihari, 45, who clearly enjoys the stir caused by his tireless agitation.

The *New York Times* journalist also interviewed some of Mr Bihari's fellow dead, among them the bedraggled Ansar Ahmed:

> Ansar Ahmed, 48, lives with his widowed mother in Madhnapar. He was recorded as dead in 1982, when his brother took control of the family's small rice paddy. Madhnapar, home to 90 families, is a place of mud-brick dwellings surrounded by open fields and scum-laden ponds. Villagers have been split by this matter of life or death in their midst, with those favouring the former position giving shelter to Mr Ahmed and his mother and those supporting the latter treating him as an invisible spectre. Recently, because of pressure from the High Court, a magistrate went to Madhnapar and, after a quick inquiry, brought Mr Ahmed back to life. Criminal charges have been filed against his brother, Nabi Sarwar Khan, who is quite grouchy about this change of fortune. 'These are only allegations,' Mr Khan said gruffly in his own defence. As various cases are investigated, several treacherous relatives and the venal officials who abetted them have likewise been charged.

As head of the Association of Dead People, Lal Bihari sought and got lots of attention. Eventually, this made it easier to gather information about his shadowy peers. He came to estimate that, in his province of Uttar Pradesh alone, there were more than 10,000 living dead.

There is a longstanding joke that, in America, in the city of Chicago, dead voters have been known to cast the deciding margin in close elections. In India, in Uttar Pradesh, the situation is just the reverse, and quite real. The living dead would, many of them, desperately love to vote. But the Indian bureaucracy is too efficient to permit that, and almost no amount of patient arguing by a democracy-loving dead person will change an Indian bureaucrat's mind.

Led by the indefatigable Bihari (of whom it might be said, 'You can't keep a good man dead'), the Association of Dead People has given up on patient arguing. They go in for noisy, flashy demonstrations, the more bureaucratically embarrassing the better. They hold political rallies. They stage public funerals for themselves. They try to get arrested, hoping to force the police to list their names on an arrest warrant, which would be, of course, an official government document acknowledging that they are alive. They have even run for political office.

And, every once in a while, it works. So far a handful of dead have been officially returned to life. Hopes are high that, with continued publicity, hundreds or thousands more will be able to step lively and join them.

But the Association of Dead People has even higher ambitions. The members hope to shame the system so much that, in future, greedy people will have a hard time doing to their relatives what Lal Bihari's uncle did to him.

For helping dead people literally get their lives back, and for giving hope to a nation benumbed by its dashpot bureaucracy, Lal Bihari was awarded the 2003 Ig Nobel Peace Prize.

Lal Bihari *almost* made it to the Ig Nobel Prize Ceremony.

The Ig Nobel Board of Governors had had a difficult time getting in touch with him. First came an elaborately arranged telephone conversation between various parties who spoke highly dissimiliar dialects of Hindi. Then there was a long period of touch-and-phone-line-go-dead communications, in bits of two languages augmented by electronic tones, between Cambridge, Massachusetts, and the one little shop in Azamgarh that had a fax machine that could sometimes get messages to Lal Bihari.

Finally, things moved into high gear. Indian film maker Satish Kaushik had read about Lal Bihari, had visited him and decided to make a movie about his life. When the Ig Nobel Prize entered the picture, Kaushik took things in hand, and offered to try to get a passport for Lal Bihari to travel to the United States, and to pay for the journey. Somehow, he succeeded. The Indian Government issued a passport to the man it had for so long insisted was dead.

But the travel plans came to naught. The United States consulate in Mumbai refused to grant Lal Bihari a visa to travel to Harvard. Apparently, being dead was the least of his problems.

Satish Kaushik arranged for his cinema colleague Madhu Kapoor to fly to the Ceremony. There she took custody of Lal Bihari's Ig Nobel Prize, handed to her by Nobel Laureate Dudley Herschbach amidst thunderous applause. Kapoor told the audience: 'For the first time in the world, living people are honouring the living dead. Lal Bihari died in 1976, but he is feeling much better.' She offered thanks on behalf of the former dead man, and promised that she would deliver the Prize to him. A month later, at a ceremony in India, Amar Singh, the general secretary of the Samajwadi Party, presented Lal Bihari with his Ig Nobel Prize.

The uncle who once consigned Lal Bihari to a life

of death is now himself deceased, officially and in all other ways. The uncle's sons now have possession of the land. Lal Bihari is content to let them keep farming there, with dirt on their hands, and guilt presumably weighing heavy on their heads.

The Birth of Murphy's Law

The universal aptitude for ineptitude makes any human accomplishment an incredible miracle.

— John Paul Stapp, medical officer and human test dummy of the rocket-sled project

The Official Citation

The Ig Nobel Engineering Prize was awarded to:

The late John Paul Stapp, the late Edward A. Murphy, Jr, and George Nichols, for jointly giving birth in 1949 to Murphy's Law, the basic engineering principle that 'If there are two or more ways to do something, and one of those ways can result in a catastrophe, someone will do it' (or, in other words: 'If anything can go wrong, it will').

IG NOBEL PRIZES 2   For a detailed history of the birth and subsequent history of Murphy's Law – and the heroic role played by John Paul Stapp – see 'The Fastest Man on Earth', by Nick T. Spark, *Annals of Improbable Research*, volume 9, number 5, September/October 2003. Much of the account here is based on Nick Spark's research. Some additional information is also available from the museum at Edwards Air Force Base.

For most people, the phrase 'Murphy's Law' is syn-
onymous with the saying 'If anything can go wrong,
it will.' Not so many know that there really was a
Murphy, and that his full name was Edward Aloysius
Murphy, Jr. Fewer still are aware of the disagreements
over who thought up the phrase 'Murphy's Law', and
over what, exactly, motivated that person.

The phrase 'Murphy's Law' was born, one way or
another, from a moment of exasperation.

After World War II, the US Army Air Force ran
some frightening tests in the desert at Muroc Air
Force Base (later renamed Edwards Air Force Base),
in California. They wanted to know how much force
a human body could endure.

This was not just to satisfy somebody's curiosity.
The question was whether more people would survive
aeroplane crashes if the planes were built sturdier.

The project engineers built a rocket sled – a metal
frame with a chair on top, wheels on the bottom, and
bottle rockets attached to the rear. The sled rode on
a long, straight stretch of railroad track. It could
accelerate to speeds well over 200 mph – and then,
thanks to special braking mechanisms, come to a
screeching halt.

John Paul Stapp was the project's medical officer.
He couldn't tolerate the idea of overseeing some test
pilot's possible death on the rocket sled. So he decided
to ride the sled himself.

The main goal was to measure how much force the
test pilot endured when the sled slammed to a halt.
But the project lacked a crucial piece of technology,
a set of electronic transducers needed to gauge that
force.

John Paul Stapp riding the rocket sled. The sled was accelerated to several hundred miles per hour, then stopped as suddenly as possible. (Photo courtesy of Edwards Air Force Base History Office)

That's why Murphy entered the picture. The engineers at Muroc Air Force Base heard that Captain Murphy, located out in Ohio, almost 2000 miles to the east, was a whiz at electronics. They got in touch, told Murphy what they needed, and Murphy built and hand-delivered a set of transducers.

A technician installed them, and the rocket sled was sent on a test run with a dummy strapped into the seat. After the test run, the engineers looked at the gauges. The readings were zero, as if the test run had not even occurred. Something had gone wrong.

Murphy examined the transducers, and immediately saw the problem. The technician had installed them backwards and upside down.

It was at that moment that Captain Edward A. Murphy, Jr, said ... whatever it was he said. Something to the effect that 'If there's any way they can do it wrong, they will.' But no one wrote it down. It was just one remark made by a very frustrated man. Much later, people would disagree as to whether Murphy had been referring to the technician who had built the transducers, or to the technician who installed them, or to some general philosophical principle.

That day, Murphy installed the transducers the correct way, watched a successful test run, and then went home. That was the end of his involvement.

Then the project went ahead full force.

John Paul Stapp rode the rocket sled repeatedly, at increasingly higher speeds followed by increasingly violent stops. He endured forces greater than 36 gs – 36 times the everyday, down-to-earth force of gravity. Until then, engineers and doctors had believed that no one could survive half that much force. And because of that mistaken belief, aeroplanes had been built too flimsily, with inadequate seat belts, and crash victims had died needlessly.

This was big news, as were the spectacular photos of John Paul Stapp riding the sled. At a press conference, somebody asked Stapp, 'How is it that no one has been severely injured – or worse – during your tests?' Stapp's reply: 'We do all our work in consideration of Murphy's Law.' He then explained that they had taken great care to think through all the possibilities *before* doing the actual test runs.

The reporters were charmed. They began telling the world about Murphy's Law. After that, word just kept on spreading.

About fifteen years passed, though, before Murphy himself first heard the phrase 'Murphy's Law'. By that time, it had become part of the language – part of many different languages, in fact.

John Paul Stapp, seen in a series of photos taken as the rocket sled came to a sudden, violent stop.

In various places at various times, credit for coining the phrase was attributed to Stapp, or to Murphy, or to George Nichols, the project's director and a close friend and admirer of John Paul Stapp. George Nichols, though, was neither friend to, nor admirer of, Murphy. According to George Nichols, the phrase 'Murphy's Law' was meant as something of a dig at Murphy.

Murphy died in 1990, Stapp in 1999. The Law endures.

For their contributions to human understanding, the late John Paul Stapp, the late Edward A. Murphy, Jr, and George Nichols were awarded the 2003 Ig Nobel Prize in the field of engineering.

John Paul Stapp's widow, Lilly, could not travel to the Ig Nobel Prize Ceremony. Historian Nick Spark delivered an acceptance speech on her behalf:

> John Paul Stapp said, 'The malignant adversity of inanimate object rivals the demonic havoc of catastrophic events in producing human frustration.' And he may or may not have been responsible for naming Murphy's Law, but from his perspective, Murphy's Law serves as a warning: Think about what might go wrong, and what you can do in advance to prevent that from happening.

EDWARD ALOYSIUS MURPHY
BENICIA, CALIFORNIA
Senatorial

Our earliest memory of Murf is of a plebe convulsed with laughter at the antics of the Beast Detail. Clutching his sense of humor, he scurried through the Academy never more than one jump ahead in academics, but never too harried to entertain with his polished imitations. Abounding with ideas, he sought new solutions for each problem, and he enjoyed nothing so much as an argument on his methods. Murf's originality amused and amazed us; his friendly grin won a place in our memory.   "*Murf*"

Sergeant (J) Hundredth Night Show (1-2).

Edward A. Murphy III accepted the Prize on behalf of his late father, who, he said, 'would've been honoured to receive this award. In fact, he would have been thrilled even to be invited to an event like this.'

George Nichols had plans to attend the ceremony, but was in ill health and instead sent a tape-recorded speech. In it, he carefully expressed the reverence that good engineers, everywhere, always have for the power of The Law. He also

Edward A. Murphy, the Murphy of Murphy's Law, as seen in the yearbook from his college, West Point, a decade before the incident that gave rise to the phrase 'Murphy's Law'.

expressed his belief that Murphy gets too much credit for Murphy's Law.

IGNOBEL PRIZES 2 Thanks to John Paul Stapp, many people are alive today who might otherwise have perished in crashes. And not just aeroplane crashes.

Stapp pointed out to the Air Force that more pilots were killed in car crashes than in plane crashes. Then he pushed long and hard to have seat belts installed in automobiles, as well as in aircraft. In 1966 the US government enacted a law forcing car manufacturers to install seat belts. This was in large part due to Stapp's lobbying. For Stapp, it had been a pure, life-and-death application of Murphy's Law: plan ahead for whatever is at all likely to go wrong.

The Further Adventures of Troy

As the fearless inventor of the Ursus series of bear suits, Troy Hurtubise has been hit by trucks, pummelled by baseball bats, cut by industrial saws, and knocked down the side of the Niagara Escarpment. Yet ensconced in the plastic and metal Ursus Mark VI, and its successor Mark VII, Hurtubise, 39, has walked away unscathed and ready for the next encounter, which once involved being run into by a 20-ton front-end loader.

— from a report by journalist Phil Novak of BayToday.ca, 26 March, 2004

The Official Citation

The Ig Nobel Safety Engineering Prize was awarded to:

Troy Hurtubise, of North Bay, Ontario, for developing and personally testing a suit of armour that is impervious to grizzly bears.

IgNOBEL PRIZES 2 For the basic story of Troy Hurtubise and his work, see the chapter called Troy and the Grizzly Bear in the book *Ig Nobel Prizes*. (Troy is, by the way, the only Ig Nobel Prize winner who has an entire chapter devoted to him.) To see and hear Troy in action, watch the documentary film *Project Grizzly*, produced by the National Film Board of Canada. Some video snippets of Troy's further and later adventures are posted on his website, http://projecttroy.com/.

For Troy Hurtubise – panner for gold in the Canadian wilderness; survivor of an unprepared encounter with a grizzly bear; designer, builder and tester of an anti-grizzly-bear suit of armour made from titanium, duct tape, hockey pads and other scrounged materials; 'test pilot' when the suit was hit by a 136-kilogramme (300-pound) tree dropped from a height of 9 metres (30 feet); 'test pilot' (again) when the suit was rammed eighteen times by a three-ton truck travelling at 50 kilometres an hour (30 mph); 'test pilot' (again) when the suit was methodically assaulted by humungous bikers armed with a splitting axe, planks and baseball bats; and survivor of an encounter with a bankruptcy court that took

Troy Hurtubise and
the Ursus Mark VI.
(Photo courtesy of
the National Film
Board of Canada)

possession of the anti-grizzly-bear suit – for Troy, the
adventure continues.

The first, famous chapters of Troy's adult life have been
widely reported. The later chapters are still taking form.
Here is a shorthand version of what Troy Hurtubise has
undertaken and endured recently.

   Troy completed work on the next-generation version
of the suit. The Mark VI, familiar to everyone who saw the
documentary film *Project Grizzly*, is primitive in contrast
to the Mark VII. Where the former has a rough-and-
unpolished look, its successor gleams. And where the
Mark VI was, by Troy's description, 'created with a 600-

pound (270-kilogramme) grizzly in mind,' the Mark VII was designed to 'withstand an attack from the larger Kodiak subspecies, members of which can weigh 1200 pounds (540 kilogrammes) and stand 10 feet (3.05 metres) upright.'

Yes, a Kodiak bear. A Kodiak is bigger than a grizzly. This design specification – able to withstand a Kodiak bear – sprang from the results of a specific incident. Troy, wearing only his Ursus Mark VI suit and perhaps some underwear, faced down two Kodiak bears. A report in Troy's local newspaper, the *North Bay* (Ontario) *Nugget*, tells what happened:

Troy Hurtubise looked so scary in his Ursus Mark VI suit that a 585-kilogramme Kodiak took 10 minutes to approach him, while a 157-kg grizzly didn't want to, period. Hurtubise, the North Bay inventor and star of *Project Grizzly*, climbed into his virtually indestructible body shell this weekend – exactly where was kept secret to keep the media at bay – because he wanted to test the suit against a real, live bear. But things turned out differently than expected, with the Kodiak, owned by an American animal trainer, avoiding Hurtubise until it could muster the courage to approach him.

The trainer, concerned about the size mismatch – the bear measures three metres when upright, and weighs 450 kg more than the besuited Troy – would not allow the Kodiak to get any closer. Although the controlled attack Hurtubise hoped for didn't take place, his suit was still bear-tested, and passed with flying colours, he said. In order to familiarize the Kodiak with the suit, the trainer gave it to him in pieces. 'He literally claimed that suit as if it was a dead

Troy Hurtubise and
the Ursus Mark VI.
(Photo courtesy of
the National Film
Board of Canada)

kill and pulled it in under his chest and started to put
his weight on the upper body part,' Hurtubise said.
'He tore some chunks of the rubber off, but even
though he was pounding on it with all 1300 pounds,
he wasn't able to crush it.'

As a result of that test of the Mark VI, Troy launched
into the design and building of the Mark VII. Although
he has not yet brought the more advanced suit up against
a Kodiak, he has tried it out against a piece of heavy construction equipment. Troy, in the Mark VII, stood against
a brick wall. A 20-ton front-end loader crashed into him,
driving Troy into and through the wall. Troy was
unharmed. Video of the test is posted on Troy's website.

Internally, the Mark VII is bristling with innovations –

among them a three-inch LCD video screen, electric cooling fans, a voice-activated two-way radio and a robotic grappling arm controlled by a tongue switch in the suit's headpiece.

But Troy has not confined his attentions to bear suits.

He also invented what he calls a 'fireproof paste', which he demonstrated in 2003 on the Discovery Channel Canada television programme *The Daily Planet*. Viewers saw Troy put on a hockey helmet that was coated with a thin layer of the paste. An assistant then applied a blowtorch to Troy's head. Hurtubise told reporters:

> I could coat the belly of the NASA space shuttle with fire paste for twenty-five thousand dollars (US), instead of the sixty million dollars it costs for them to put tiles on it.
> It can stand up to the heat of re-entry to the earth's atmosphere, and then they can simply wash it off.

In early 2004, Troy was accorded one of modernity's singular honours. An entire episode of the television programme *The Simpsons* paid tribute to him, with Homer Simpson re-enacting rather tame versions of things that Troy had done for real.

It is too early for anyone to say, with certitude, where this will lead. Troy has big plans. He also has a big difficulty: how to fund his ongoing programme of invention and testing. He has considered whether to put the suits – the Mark VI and the Mark VII – up for auction on eBay. The Mark VI is now owned by the bankruptcy court, but the court has a willingness and encouragement for Troy to work something out.

When Troy was awarded the 1998 Ig Nobel Prize in the field of Safety Engineering, he came to the Ig Nobel

Ceremony and triumphantly accepted his Prize. At the time, as since, people laughed in amazement. Most of them also smiled in appreciation of Troy's most impressive quality. Though some see him as half-genius and others as half-crackpot, all must acknowledge that Troy is very, very careful. The proof is that he is still, after all these technologically innovative adventures, alive.

2. The Little Things in Life

It's the little things that matter. Or so people say. Here are three little things that led to Ig Nobel Prizes:

You Can Rent Liechtenstein

Politicians' Uniquely Simple Personalities

Trinkaus – An Informal Look

You Can Rent Liechtenstein

RENT A STATE IN LIECHTENSTEIN

After researching, talking, negotiating and adapting things we finally signed the contract in January with Liechtenstein Tourism to start the world's first state rental for Corporate events in Liechtenstein. The media information started off big interest from all kinds of media. Reuters, BBC, CNN, the *Sunday Times*, the *Guardian*, *Newsweek* and *Travel Weekly* were the most prominent media to be interested by this new Meetings, Incentives, Conventions and Events concept.

Details about Rent a State will be available in April on the renewed Xnet Rent-a-Village website, www.rentavillage.com.

We will be pleased to answer your inquiries or special requests for this concept. If you like to join a Fam Trip to Liechtenstein from March 21st to March 23rd 2003, please register on our website and we will inform you about the trip. Please contact Karl Schwärzler for further information under schwaerzler@xnet.li or mobile phone 00423 79 11919.

— detail from a press release issued 21 May, 2003

The Official Citation

The Ig Nobel Economics Prize was awarded to:

Karl Schwärzler and the nation of Liechtenstein, for making it possible to rent the entire country for corporate conventions, weddings, bar mitzvahs, and other gatherings.

> Nestled in the tree-lined scalps
> Of the mountains called 'The Alps'
> There's a nation that's divine.
> Yes, of course! It's Liechtenstein!
>
> Each and every resident,
> More or less, is up for rent.

Fully furnished, rain or shine!
Weekly leases! Liechtenstein!

There are bad landlords and there are good land-lords. The general expectation is that Liechtenstein – the Principality of Liechtenstein – will be a good land-lord. The place is well kept, the neighbourhood is nice. Those nearby seem to get along happily with their petite, self-sufficient neighbour.

In 2003, the tiny nation began making preparations to rent itself out, with help from a company called Xnet, which had for several years been arranging for individual alpine villages to rent themselves out. Karl Schwärzler, Xnet's director, told reporters that 'This will be good for the country. It has an image of banks and "special" finance houses, but this will allow people to see a different side of Liechtenstein.'

Liechtenstein. Photo: from the official Liechtenstein Tourism website.

The British newspaper the *Guardian* interviewed a Danish gentleman named Jeppe Weinreich, who spends part of every year working as a waiter in little Liechtenstein:

> 'It's a funny thing. But this is a strange country. There are only 324 unemployed people; it can fit into the territory of Luxembourg 12 times over; you can see the whole country in two hours, and the local prison can only hold 20 prisoners. So why not rent it out? It's a bit boring, but it's not dangerous to walk the streets at night – and it's rich,' adds Mr Weinreich. 'Seasonal workers in other countries come from places like Albania, but here they come from Scandinavia.'

Even so, not everything or everyone will be for hire under Liechtenstein's new venture. The services of the country's reigning monarch, Prince

Hans-Adam II, cannot be purchased. Not u
lacks a mercantile instinct himself. Furious th
the country's politicians are reluctant to give h
more power, he is threatening to sell the royal
castle to the Microsoft billionaire Bill Gates – or
anyone else who will buy it – and move to Vienna.

Liechtenstein has been an independent nation since
1806. It was neutral in World War I and World War II.
You cannot rent Liechtenstein's army, because it does
not have one. Yet the options offered to a potential
renter are numerous and of many kinds. There are lakes,
mountains, and two castles. There are eleven villages,
and lots of cows. There is quiet, and a soothing lack of
glitter.

You can arrange for 'Night Tobogganing with Après
Ski and a Cheese Dumpling Party'. You can bike cross-
country – all the way from Austria to Switzerland and
back, all in the same day. Several times in that day, if
you like that sort of thing.

You can romp in the Prince's Winery, and go on a
treasure hunt through his vineyards. See the castles.
Take in the Bird of Prey Show (and even arrange to
coordinate it with your corporate product
presentations). Go down into an old mine, and come up
again. The official brochure offers many other delights,
including (and I quote):

- Standardized accommodation with local hosts in
  small hotels and guest houses
- Key raffle to find your host for the next couple of
  days
- Symbolic and ceremonial key handover through the
  mayor joined by the local brass band and folklore
  groups

Fürstentum Liechtenstein

Program Highlights:

Wine Tasting in the Prince's Winery
Tast the Prince's Wines and go on a treasure hunt throuh his vinyards

Castle Gutenberg
A medieval Castle in Balzer with a view over most of the country.
A very good location for unforgettable feasts and cultural highlights.

Sunset 1000m above the Rhine Valley
Walking on a trail through steep rocks with a fantastic view, continue in
a valley with no civilization or have a torch walk back.

Bird of Prey Show for Product Presentations

The Gonzen Mines
A Journey to Middle Earth with obstacles to overcome.
Suitable for Leadership Trainings and Conflict Management Trainings.

Igloo Meetings
Igloo Village for Break Out Rooms

Liechtenstein Museum of Art
Use this extraordinary location for your meetings and conferences
between Rubens and Rembrandt Paintings.

Labyrinth of Corn
Make a Treasure hunt in a labyrinth of corn in the vast fields
of the Rhine Valley.

Xnet AG . Im Bartledura 14 . FL-9494 Schaan . T +423 2301696 . F +423 2301695 . info@xnet.li

A promotional flyer that lists some cosy charms of the rental property.

- 🏆 Regional outdoor and indoor training possibilities according to the wishes of the company (a day in a winery, smuggling tours, etc.)
- 🏆 Unusual locations as convention facilities (chalets, igloos, museum or castles)
- 🏆 Tailor-made marquee solutions for the evening events, conventions, and meals (depends on the size of the village, starting from 250 people upwards)
- 🏆 Decoration and branding of the village: renaming streets according to products or even change the village name into your brand

Yes, you can 'brand' Liechtenstein, as the marketing world would put it. You cannot, however, 'rent to own'. Nor (to answer the question that everyone always asks) can you rent the women or men of Liechtenstein. At least, that is not part of the official offer.

For their innovative contribution to cosy international accommodation and hospitality, Karl Schwärzler and the nation of Liechtenstein shared the 2003 Ig Nobel Prize in the field of Economics.

Karl Schwärzler travelled to the Ig Nobel Prize Ceremony. After humbly accepting the Prize, he stepped to the lectern and said:

> Thank you very much, on behalf of my colleagues in the nano-country of Liechtenstein. This is about everything you see there (Mr Schwärzler gestured to the images of Liechtenstein that were being projected above the stage). It's only 61 square miles, and we have 80 per cent mountains. It's a beautiful country, and, uh, yeah, you can rent it. Thank you very much.

Ig Nobel Economics Prize winner Karl Schwärzler explains how simple it is for anyone to rent Liechenstein. (John Bradley/*Annals of Improbable Research*)

After the Ceremony, bubbly audience members approached him to ask about rental rates and about whether Liechtenstein allows tenants to have pets. Standing nearby, several students from the Harvard Business School listened intently and took careful notes.

Politicians' Uniquely Simple Personalities

Call it the 'Arnold Effect'. The straight-talking Hollywood action star's election win in California has had an electrifying impact on Germany, leading to calls on Friday for top politicians to voice clear ideas in simple language or be swept away at the polls. 'The more confused we are by what they say, the greater our longing for a man or woman with simple words,' wrote *Bild* newspaper columnist Franz Josef Wagner. 'The only problem is that it's the wrong ones who usually master simple language.' Schwarzenegger's victory in the California race for governor has led to editorials calling for German politicians to abandon their barely comprehensible speaking style in favour of 'Klartext' (straight talk).

— from a report by Reuters, 16 October 2003

The Official Citation

The Ig Nobel Literature Prize was awarded to:

Gian Vittorio Caprara and Claudio Barbaranelli of the University of Rome, and Philip Zimbardo of Stanford University, for their discerning report 'Politicians' Uniquely Simple Personalities'.

IG NOBEL PRIZES 2 Their study was published in *Nature*, volume 385, 6 February 1997, p. 493.

Scientists try to make sense of a complicated phenomenon by boiling it down into a few numbers. Voters try to make sense of a politician by doing much the same thing.

Even the simplest person has a uniquely complicated personality. Yet most of us manage to size up other people's personalities. We do it all the time, without great difficulty. Psychologists think they know, roughly, *how* we do it. For practical purposes, almost anyone's personality can be boiled down to a mere five aspects. The best-known version of the 'Big Five' theory lists these five factors:

- energy/extroversion
- agreeable/friendliness
- conscientiousness
- emotional stability against neuroticism
- intellect/openness to experience

Now, precisely defining the five personality aspects can be tricky – psychologists argue over that all the

Politicians' uniquely simple personalities

The complexity of human personality has been reduced to five dimensions, based on factor analyses of judgements of personality traits[1]. Many researchers agree that a five-factor model of personality cap-

son's one vote: be it for or against. Therefore, we predicted that personality judgements about political candidates would likewise be constricted to involve a limited number of factors rather than the usual

The first of the two stable[6] personality factors for politicians has been named energy/innovation (which is a blend of energy and openness), and the second factor is honesty/trustworthiness (a blend

Caprara, Barbaranelli and Zimbardo's Prize-winning report.

time. But the important thing is that just *five* seems to be plenty. You don't need ten, or twenty, or twenty million. Just five.

Three psychologists – Gian Vittorio Caprara, Claudio Barbaranelli and Philip Zimbardo – wondered whether it might be different when people judge a politican. After all, feelings about a politician eventually come down to just *two* choices: either 'Yes, I would vote for this person' or 'No, I wouldn't'.

So, these scientists wondered, maybe people do a little extra boiling down when they judge a politican. Maybe, for practical purposes, a politician's personality is simply twofold.

When these scientists tested their idea, that's exactly what they found.

Caprara, Barbaranelli and Zimbardo studied more than 2,000 Italian citizens, asking each of them to make judgments about some famous individuals: 'Ratings were made of two Italian political candidates (Silvio Berlusconi and Romano Prodi), an international celebrity (skiing hero Alberto Tomba) and a famous Italian personality (Pippo Baudo).' The Italian citizens were also asked to make judgments about themselves.

The results of the study:

When people judged their own personalities or those of TV stars or sports heroes, they boiled it down to five aspects. But when they judged a

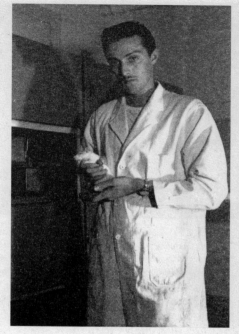

A photograph, taken circa 1956, of Philip Zimbardo and a lab rat. Zimbardo went on to a career as one of the world's great psychologists. (Photo courtesy of George Slavich and Philip Zimbardo)

politician, it came down to just two things – how energetic is he (or she), and how trustworthy?

Later, the scientists repeated their investigation, but in a different country. They asked a bunch of American citizens to rate their own personalities, and also to rate the personalities of Presidential candidates Bill Clinton and Bob Dole, and of basketball star Ervin 'Magic' Johnson. The results were about the same as in Italy: politicians had simpler personalities than other people.

Here is the scientists' published conclusion, in official technical lingo:

We conclude that, by adopting a simplifying method

Ig Nobel Psychology Prize winner Phil Zimbardo bursts through the curtain as he is announced. Nobel Laureate Rich Roberts, (right, wearing moose-antler hat) waits to hand him the Prize. Ig Nobel attendant Tom Ulrich holds the curtain. (Margaret Hart/*Annals of Improbable Research*)

of judging political candidates' personalities, voters use a cognitively efficient strategy for coding the mass of complex data, thus combating information overload. Doing so helps them to decide how to vote.

For helping us understand how we understand politicians, Gian Vittorio Caprara, Claudio Barbaranelli and Philip Zimbardo were awarded the 2003 Ig Nobel Prize in the field of Psychology.

Philip Zimbardo journeyed to the Ig Nobel Ceremony, where he burst onto the stage with an energy and enthusiasm far beyond that of most other 70-year-olds. That same day, the Associated Press interviewed him: 'Zimbardo said the research could be useful to campaign strategists. "This is a very strange thing, this Ig

Nobel. I had heard vaguely about it. At first, I thought it was an insult," Zimbardo said. Now he's honoured, he said, and hopes the publicity will build interest in his research.'

Later, Zimbardo got down to business with a reporter from his home state of California: 'Politicians like to think of themselves as so complex, but the electorate thinks of them as simple.'

During the next several weeks, many people noted, with varying mixtures of delight and other emotions, a happy coincidence. The citizens of California took part in a hastily called election to recall, and perhaps replace, their state's governor. On Tuesday, 7 October, more than five years and nine months after the publication of Caprara, Barbaranelli and Zimbardo's scientific report 'Politicians' Uniquely Simple Personalities', actor Arnold Schwarzenegger was elected to be the 38th governor of the State of California.

Trinkaus – An Informal Look

Investigated the use of the noun 'integrity' in social conversation. A counting of its employment during 67 half-hour television talk shows showed 7 instances. The author concludes that the use of the word is fading.

— from the report 'Conversational Usage of "Integrity": an Informal Look,' by John W. Trinkaus, *Perceptual and Motor Skills*, volume 86, number 2, April 1998, pp. 585–6

John W. Trinkaus prepares to accept his Ig Nobel Prize from Nobel Laureate
Wolfgang Ketterle. The Prize – a bar of solid gold one nanometre long – is
somewhere inside the transparent plastic box Professor Ketterle is holding.
(John Bradley/*Annals of Improbable Research*)

The Official Citation

The Ig Nobel Literature Prize was awarded to:

John Trinkaus, of the Zicklin School of Business, New
York City, for meticulously collecting data and
publishing more than 80 detailed academic reports
about things that annoyed him, such as: what
percentage of young people wear baseball caps with the
peak facing to the rear rather than to the front; what
percentage of pedestrians wear sport shoes that are
white rather than some other colour; what percentage
of swimmers swim laps in the shallow end of a pool rather
than the deep end; what percentage of automobile
drivers almost, but not completely, come to a stop at
one particular stop sign; what percentage of commuters

carry attaché cases; what percentage of shoppers exceed the number of items permitted in a supermarket's express checkout lane; and what percentage of students dislike the taste of Brussels sprouts.

IG NOBEL PRIZES 2 Nearly all of Professor Trinkaus's Prize-winning reports appeared either in the journal *Psychological Reports* or in its sister publication, *Perceptual and Motor Skills*. For a list of 86 of Professor Trinkaus's publications and an appreciation of his work, see Trinkaus: an Informal Look, *Annals of Improbable Research*, volume 9, number 3, May/Jun 2003, pp. 4–15.

Does one man count for nothing?

The man is John W. Trinkaus, a professor emeritus at the Zicklin School of Business in New York City. When something sufficiently annoys Professor Trinkaus, he takes the time and trouble to carefully count that something – how numerous it is or how often it happens. And then he publishes a report.

Trinkaus's published reports are numerous (more than 90) and pithy (with few exceptions, each is a page or two in length). They concern a wide range of common behaviour that struck Dr Trinkaus as curious or grating.

It all began when he was called to jury duty, and noticed that not all of his fellow jurors were eager to be there. This led to the first of Professor Trinkaus's many 'Informal Look' reports in 1978: 'Jury Service: an

Informal Look'. In it, Professor Trinkaus concluded that 'potential jurors are divided into those who do and those who do not want to serve'.

Since then, Trinkaus has conducted research on little-noticed aspects of parking lots, on the marital status of television quiz-show contestants, and on many other things.

What percentage of shoppers in a supermarket's express checkout lane have more than the number of items permitted? For the answer, see Trinkaus's 1993 paper 'Compliance With the Item Limit of the Food Supermarket Express Checkout Lane: an Informal Look'. How did it change during the irrational exuberance of the dot-com years? See his 2002 paper 'Compliance With the Item Limit of the Food Supermarket Express Checkout Lane: another Look'.

After a snowstorm, what percentage of drivers don't bother to clear the snow off their car roofs? See 'Snow on Motor Vehicle Roofs: An Informal Look' (2003).

What percentage of people leaving a building choose a door that is already open, rather than one that is closed? See a pair of reports from 1990: 'Exiting a Building: An Informal Look' and 'Exiting: another Look'.

Trinkaus has collected and published data on those phenomena and oh, so many more.

Trinkaus reports the numbers. Period. His research, he says, is purposely simple. 'It suggests numbers for occurrences which many times folks talk about just in qualitative terms. I'm suggesting that people don't always have to settle for "quite a bit", "a lot", "a bunch".'

For example, his 1991 paper 'Taste Preference for Brussels Sprouts: an Informal Look' delivers crisp confirmation: 54% of young students found the vegetable to be 'very repulsive'.

STOP SIGN COMPLIANCE: AN INFORMAL LOOK

JOHN TRINKAUS

Baruch College, City University of New York

Reason (1974) reports that reckless driving is a behavior pattern con-
ating significantly to road accidents. Ugwuegbu (1977), in a field study
esponses of drivers to stop signals, observed about a 40% adherence rate.
To determine the trend of this action form, in particular as it relates to

STOP SIGN COMPLIANCE: ANOTHER LOOK

JOHN TRINKAUS

Baruch College, City University of New York[a]

Trinkaus' (1982) informal three-year inquiry implies that drivers' com-
ance with stop signs is decreasing. To determine the possible presence of
continuance of this trend, the study was replicated for an additional two years.
1 test conditions were the same: the setting (a residential suburban com-

STOP SIGN COMPLIANCE: A FURTHER LOOK[1]

JOHN TRINKAUS

Baruch College

Trinkaus reported a steady decline, over a 5-yr period, in driver compliance with stop
s (3, 4). To gain some insight as to whether this trend continued, the inquiry was replicated
an additional 4 yr.
Observations were made at the same four T-junction intersections in a residential suburban
imunity in the Metropolitan New York area. The primary purpose of the signs is to discour-
the flow of through traffic on the local streets. Convenience citings were again conducted
ing 90-min. intervals, in the morning and evening hours, of the four days immediately after
or Day. As before, no note was made of the type of vehicle, characteristics of the occu-
ts, or the weather conditions.

Year	Total No. Occurrences Observed	Distribution of Responses						
		Full Stop		Rolling Stop		No Stop		
		f	%	f	%	f	%	
1984	287	32	11	57	20	198	69	
1985	303	33	11	52	17	218	72	

STOP SIGN COMPLIANCE: A FOLLOW-UP LOOK[1]

JOHN TRINKAUS

Baruch College

Summary.—Analysis of 324 observations at a previously studied intersection
showed a continuing decrease in full stops (n = 13) with increased rolling (n = 19) and
no stops (n = 292), suggesting change in meaning of beliefs.

Based on simple informal inquiries conducted once a year for nine consecutive years (19

TASTE PREFERENCE FOR BRUSSELS SPROUTS: AN INFORMAL LOOK[1]

JOHN TRINKAUS AND KAREN DENNIS

Baruch College

Summary.—An inquiry of the taste preference of 442 business students for brussels sprouts showed about a 30% dislike of the vegetable, 40% Indifference, and a
10% like. Some implications of the findings are suggested.

COMPLIANCE WITH THE ITEM LIMIT OF THE FOOD SUPERMARKET EXPRESS CHECKOUT LANE: AN INFORMAL LOOK[1]

JOHN TRINKAUS

Summary.—A total of 75 15-min. observations of customers' behavior at a food
supermarket suggests that only about 15% of shoppers observe the item limit of the
express lane. The overages tend to be limited to approximately 1 to 3 pieces.

COLOR PREFERENCE IN SPORT SHOES: AN INFORMAL LOOK[1]

JOHN TRINKAUS

Baruch College

Summary.—Observations on five workdays at a large terminus for a number of com-
muter rail lines indicated that among 4731 passengers 5% of 2794 men and 31% of
1937 women wore white sport shoes. Questions relative to interpretation are raised.

STOP SIGN COMPLIANCE: A FINAL LOOK[1]

JOHN TRINKAUS

College of Business Administration
St. John's University

Summary.—A concluding study to a number of informal enquiries conducted
over the years, on drivers' compliance with stop signs in a residential community, suggests that this traffic control device is now largely ignored by motorists. During a 17-
yr. period full stops declined from about 37% to 1% and rolling stops from approximately 34% to 2%.

Based on simple informal enquiries conducted once a year for nine con-
ecutive years (1979 through 1987), and another viewing five years later

Portions of a few of Professor Trinkaus's Ig Nobel Prize-winning reports.

1983 saw the publication of 'Human Communications: An Informal Look', in which Trinkaus
explains that he 'studied whether 750 riders of low-
speed self-service elevators were inclined to respond
with short utterances when one of two questions was
asked of them: 'Is this car going up?' Or 'Is this car
going down?'

The very titles of his papers stimulate thought:

'Stop Sign Compliance: An Informal Look', 1982
'Stop Sign Compliance: Another Look', 1983
'Waiting Times in Physicians' Offices: An Informal
Look', 1985

> 'Stop Sign Compliance: A Follow-up Look', 1993
> 'Stop Sign Compliance: A Final Look', 1997
> 'An Informal Look at Use of Bakery Department
> Tongs and Tissues', 1998
> 'Stop Sign Dissenters: An Informal Look', 1999

In his many traffic-related studies, Trinkaus presents, but does not attempt to explain, his most-often-observed fact: that women in vans are often the least law-abiding of drivers.

Trinkaus's eye scans in many directions. 'I just look and something strikes me. I don't purposely go out and look for topics. I just say "Ahhh – that looks good."'

> 'Wearing Baseball-Type Caps: An Informal Look',
> 1994: 'Observed 407 people wearing baseball-type
> caps with the peak in back in the downtown area and
> on two college campuses (one in an inner borough
> and one in an outer borough) of a large city. About 40
> per cent of subjects in the downtown area and at the
> inner-borough college wore the cap with the peak to
> the rear, while about 10 per cent of the outer-
> borough college subjects had the peak to the rear.'

> 'The Demise of "Yes": an Informal Look', (1997):
> 'For affirmative responses to simple interrogatories,
> the use of "absolutely" and "exactly" may be
> becoming more socially frequent than "yes". A
> counting of positive replies to 419 questions on
> several TV networks showed 249 answers of
> "absolutely", 117 "exactly", and 53 of "yes".'

While Trinkaus sometimes suggests possible ways to look at his findings, overinterpretation, or even interpretation, is something he mostly – and proudly –

avoids. When John W. Trinkaus notices something that can be tallied, he tallies it. In a profession ruled by the famous dictum 'publish or perish', Trinkaus counts.

For his carefully appreciating the little things in life, Professor Trinkaus won the 2003 Ig Nobel Prize in the field of Literature.

Professor Trinkaus gleefully journeyed to the Ig Nobel Prize Ceremony. At the age of 79, he was at last greeted by a world eager to hear more about his discoveries on broccoli-related attitudes, baseball-cap-wearing trends, and the stop-light-related behaviour of women driving vans.

Professor Trinkaus came prepared to say plenty, and the audience at Sanders Theatre, impressed, curious, and only slightly baffled, was clearly eager to listen. However, eight-year-old Miss Sweetie Poo decided, after the traditional sixty seconds had elapsed, that she had heard enough. She asked Professor Trinkaus to 'Please stop. I'm bored. Please stop. I'm bored. Please stop. I'm bored. Please stop. I'm bored. . .' Professor Trinkaus, forewarned, had come forearmed. He attempted to bribe Miss Sweetie Poo with a lollipop. Miss Sweetie Poo took the offer in her stride, accepting the sweet with a polite 'Thank you', then immediately continuing her thoughtful observation, 'Please stop. I'm bored. Please stop. I'm bored. . .' After a brief struggle, Professor Trinkaus surrendered.

3. Peace and Harmony Between the Species

Human beings share the planet with a seemingly uncountable number of other species. Relations are not always harmonious. This chapter describes a trio of Ig Nobel Prize-winning technical breakthroughs that have brought some peace and ease to the struggles for coexistence:

From Dogs' Mouths to Your Ears

Persuading Pigeons to Go Elsewhere

Chickens Prefer Beautiful Humans

From Dogs' Mouths to Your Ears

Kimiko Fukuda always wondered what her Chihauhua was trying to say. Whenever she tried to put on make-up, the small dog would pull on her sleeve. Now, she said, she knows. When her dog barks, she glances at a palm-sized electronic gadget. On its screen comes the 'human' translation: 'Please take me with you.' 'I realized that's how he was feeling,' says Fukuda, who lives outside Tokyo.

— Bow-Lingual user Kimiko Fukuda, interviewed by the *Washington Post*, 14 August 2003

The Official Citation

The Ig Nobel Peace Prize was awarded to:

Keita Sato, President of Takara Company, Dr Matsumi Suzuki, President of Japan Acoustic Lab and Dr Norio Kogure, Executive Director, Kogure Veterinary Hospital, for promoting peace and harmony between the species by inventing Bow-Lingual, a computer-based automatic dog-to-human language translation device.

 Copious technical information is available on the Bow-Lingual website http://www. takaratoys.co.jp/bowlingual/index.html.

Dr Kogure describes much of his research in a book called *Boku (Inu) no Subete wo Oshieru Wan* (*I, Dog, Will Tell You Everything About Myself, Woof!*), 2003, Field Y Publishing.

Dr Suzuki describes the development of Bow-Lingual in his book, *Bauringaru: Hajimete Inu to Hanashita Hi* (*Bow-Lingual: The Day I First Talked to Dogs*), 2003, Takeshoubo Publishing.

There is an old, old saying that a dog is a man's best friend. To whatever degree that is accurate, such friendships exist despite very strained communications. Dogs seem to understand humans fairly well – at least as well as they want to. But most humans understand dog language little better than they would understand spoken Sumerian. In 2002, a semi-secret technological project was revealed to the public. Scientists and engineers from three organizations had teamed up with some unstintingly vocal dogs. After years of study, toil, and testing, they gave the world something undreamt of by philosophers of yore: Bow-Lingual, an electronic device that translates dog barks into modern Japanese.

Bow-Lingual did not, of course, simply materialize from a void. There was a threefold creation process.

Dr Matsumi Suzuki had spent much of his career analysing recorded sounds of many kinds, including those made by various animals. After long years of performing various types of electro-acoustic analysis, he developed a set of theories about how to get at the meaning of dolphin sounds. The dog-bark research grew indirectly from that. Dr Suzuki is a legendary figure in other realms of acoustical endeavour, too. It was he who recreated the trumpeting of the woolly mammoth. It was he who assisted in some of Asia's most white-knuckled criminal investigations – among them the Benigno Aquino assassination in the Philippines, the Korean Airlines bombing, the Aum Shinrikyo subway poison-gas incident, and the Glico/Morinaga chocolate poisonings (in which the notorious 'Man of 21 Masks' is said to have sent him a taunting note that read 'Hey Suzuki, the Sound Man, do a good job of the scientific analysis!').

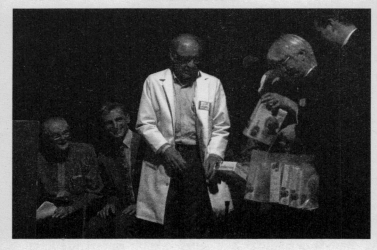

Making a return visit to the Ig Nobel Prize Ceremony in 2003, a year after he and colleagues won the Peace Prize, Masahiko Kajita of the Takara Company presents Bow-Linguals to Nobel Laureates (left to right) William Lipscomb, Wolfgang Ketterle, Dudley Herschbach and (wearing the green moose-antlers hat) Rich Roberts. (Margaret Hart/*Annals of Improbable Research*)

The second part of the team consisted of Dr Norio Kogure, president of the Kogure Veterinary Hospital, who brought many decades of expertise in understanding dog behaviour.

The third and largest chunk of the team dealt with engineering and consumer testing (testing both human consumers and dog consumers) at the Takara Company and its technical affiliate, Index Corporation. That group was led by Bow-Lingual product developer (in Japanese, his title is '*bauringaru no kaihatsu tantou*') Masahiko Kajita and overseen by Takara president Keita Sato. Kajita, a dashing bleached-blond ball of energy, seemed at times to mentally adopt a frisky-canine-from-birth perspective, as he and his colleagues and the dogs barked, listened, analysed, engineered and tested their way towards their peculiar place in history.

Translating dog language into Japanese, no simple task on the face of it, is even more complex than one might expect. There are, after all, different breeds of dog. Linguists are not at all sure that any two breeds speak exactly the same dialect, and so the Bow-Lingual inventors went to the trouble of designing and testing several variations of their circuitry, all of which they crammed into a single, tiny machine. Thus a Bow-Lingual user is able to adjust the device to translate from any of the following dialects: Airedale Terrier; Akita; Alaskan Malamute; American Cocker Spaniel; American Staffordshire Terrier; Australian Cattle Dog; Australian Shepherd; Basenji; Basset Hound; Beagle; Bernese Mountain Dog; Bichon Frise; Bloodhound; Border Collie; Boston Terrier; Bouvier des Flandres; Boxer; Brittany; Brussels Griffon; Bull Terrier; Bulldog; Bull Mastiff; Cairn Terrier; Cavalier King Charles Spaniel; Chesapeake Bay Retriever; Chihuahua; Chinese Crested; Chinese Shar-Pei; Chow Chow; Collie; Dachshund; Dalmatian; Doberman Pinscher; English Cocker Spaniel; English Springer Spaniel; Fox Terrier; French Bulldog; German Shepherd; German Shorthaired Pointer; German Wirehaired Pointer; Giant Schnauzer; Golden Retriever; Great Dane; Great Pyrenees; Havanese; Irish Setter; Italian Greyhound; Jack Russell Terrier; Japanese Chin; Labrador Retriever; Lhasa Apso; Maltese; Mastiff; Miniature Pinscher; Miniature Schnauzer; Newfoundland; Old English Sheepdog; Papillon; Pekingese; Pomeranian; Poodle (Miniature); Poodle (Standard); Poodle (Toy); Portuguese Water Dog; Pug; Rhodesian Ridgeback; Rottweiler; Saint

Bernard; Samoyed; Schipperke; Scottish Terrier; Shetland Sheepdog; Shiba Inu; Shih-tzu; Siberian Husky; Silky Terrier; Soft Coated Wheaten Terrier; Vizsla; Weimaraner; Welsh Corgi (Cardigan); Welsh Corgi (Pembroke); West Highland White Terrier; Whippet; and Yorkshire Terrier. Owners of other breeds may have to settle for an occasional mistranslation.

For inventing the world's first hand-held, computer-based, automatic dog-language-to-human-language translation machine, and thus inviting peace and harmony between the species, Keita Sato, Dr Matsumi Suzuki, Dr Norio Kogure, Masahiko Kajita and their colleagues were awarded the 2002 Ig Nobel Prize in the field of Peace.

Three of the co-inventors travelled to the Ig Nobel Prize Ceremony, accompanied by a Bow-Lingual, several sets of backup batteries (just in case!), a human translator (to translate from Japanese into English), and a man in a dog suit. The latter turned out to be Dr Suzuki's son. Upon inquiry – and people really did want to know – Dr Suzuki and the other inventors all calmly insisted that young Mr Suzuki had not been one of the dogs used in the lengthy testing process.

Two days after the Ceremony, Mr Kajita and Dr Suzuki gave a public talk at the Massachusetts Institute of Technology. Although they tried, in their speeches, to concentrate on the technical aspects of their work, the mathematicians, engineers and biologists in the audience all seemed to have one, much larger question in mind. When, they demanded, will a dog-to-English version be available?

News of this Ig Nobel Prize touched off a worldwide wave of attention for Bow-Lingual. Dog owners strained at their metaphorical leashes, rabid to get hold of it. The factory was completely overwhelmed as demand far outgrew anything the inventors or their canine collaborators had anticipated. The following year saw the arrival of a Bow-Lingual that translates dog into Korean, and another that translates dog into English. *Time* magazine named Bow-Lingual 'Best Invention of the Year'. Citizens of many nations clamoured for versions that would translate dog into French, Spanish, Chinese, Arabic, German, Hindi and virtually every other language. A lot of people, in a lot of places, began paying more attention to their dogs.

Then a second semi-secret project was revealed: Meow-Lingual, a cat-to-human translator. The arrival of the feline translator led to a new, albeit minor, headache for the inventors – demands, some serious, others less so, to produce translation devices geared to this, that, or another kind of mammal, bird, reptile, fish, or insect. The world, and perhaps the inventors, are unsure as to exactly which of these devices will arrive in the near future.

Persuading Pigeons to Go Elsewhere

Back in 1988, when we did some maintenance of the pedestal beneath the statue of the mythological warrior Yamato Takeru no Mikoto, we were amazed to find there were no bird droppings at all. At the time we first announced our discovery nobody wanted to assist us. But we determined through analysis that there seemed to be something in the composition of the statue that kept the birds away.

— Professor Yukio Hirose, interviewed in the magazine *Shukan Shincho*, 29 May 2003

The Official Citation

The Ig Nobel Chemistry Prize was awarded to:

Yukio Hirose of Kanazawa University, for his chemical investigation of a bronze statue, in the city of Kanazawa, that fails to attract pigeons.

IGNOBEL PRIZES 2 Professor Hirose has not yet published a formal academic paper about this research, but details are available from his office at Kanazawa University and from numerous press accounts [see, for example, a report broadcast on All Nippon News Network (ANN) television in May 2003].

As the heavens inevitably cover every mountain peak with snow, so do pigeons unstoppably deposit a protective white layer atop every outdoor statue – or so everyone believed. Yukio Hirose shocked and delighted the world by disproving this supposedly eternal truth.

Many people talk to pigeons, but few speak their language. Chemistry, it turns out, may be a far more effective way to communicate with birds than Japanese or English or Spanish or Chinese, or any of the other ways that people like to jabber at each other. Yukio Hirose figured this out after he noticed that *something* had consistently gotten the attention of one particular group of pigeons.

In the Kenroku garden in Kanazawa, there is a remarkable statue of the legendary hero Yamato Takeru no Mikoto. There are many things to admire about the statue, but, as a scientist, Professor Hirose was fascinated by how pristine the figure is. Unlike almost every statue in the entire history of civilization, this one is rarely visited by birds, and has seldom been given the kind of personal gifts that birds often lavish on anything that catches their fancy.

The statue is old, and the historical records hold few technical details of its manufacture. There was no obvious reason why it should stand cleanly removed from its fellows in the vast, international population of statues.

Professor Hirose analysed a small sample of the metal. Its composition turns out to be unusual. The alloy contains copper and lead, which are not uncommon in statues – but also another element that seems very out of place. The statue's old bronze is laced with arsenic.

Arsenic by itself, of course, is a poison. But when arsenic is bound up in an alloy of lead and copper, is it still somehow able to act poisonously or repellently on creatures that come near it? The answer to that question was not at all clear, and so Professor Hirose did some experiments.

He carefully prepared some new bronze, with a chemical composition very like that in the statue. He forged sheets of this metal, and allowed birds to come and pay their customary kind of courtesy visit.

This was a starkly revealing experiment. Birds consistently declined to spend time on the metal sheets, or even to come near them. Thus, concluded Professor Hirose, the statue's secret power was no longer a secret. It was simply a matter of chemistry.

Since that time he has been conducting further experiments. His hope – shared by millions of people who love statues (or at least love spending time near statues) – is that this discovery will change the world.

He is developing a technology that, if it is perfected, will give humanity a simple way to protect its statues from pigeons, crows, and other winged would-be loiterers. And to do so in a way that will not cause harm to the birds.

For his careful examinations of birds and bronze, and for his imaginative experiments with them, and for bringing hope to a problem that seemed hopeless, Yukio Hirose was awarded the 2003 Ig Nobel Prize in the field of Chemistry.

Professor Hirose and his wife travelled to the Ig Nobel Prize Ceremony. His acceptance speech was as selfless as his deeds. 'Especially,' he said, 'I thank the birds of Kanazawa, where I did this, because they gave me the idea for my research.'

As he stood upon the stage inside Sanders Theatre, humbly savouring the applause, admiration, and congratulations of 1200 audience members and four Nobel Laureates, the professor could not help but notice that Sanders Theatre contains two large statues of historical figures dressed in togas. (You can see one of these statues in the photograph on page 20.) These ancient sculpted gentlemen look down, one from either side of the stage, at the people and events passing before them. Many great and celebrated political leaders, scholars, and artists have appeared in Sanders Theatre, but in the 150-year history of the building, few have given a thought as to how very clean those two statues are. Of course they are clean. They are indoor statues, sheltered like house cats from the harshness of the great outdoors. In statuary society, they are the privileged elite, unaware of the dirty, hard days and nights endured by their less fortunate brother and sister statues.

Yukio Hirose, the 2003 Ig Nobel Chemistry Prize winner, surveys the audience as William Lipscomb, the 1976 Nobel Chemistry Prize winner, prepares to hand him the Ig Nobel Prize. Professor Hirose has brought gleaming hope to statue lovers the world over. (John Bradley/*Annals of Improbable Research*)

Soon, thanks to the curiosity and diligent work of a determined scientist named Hirose, those outdoor statues, too, may be able to hold their brows high, without fear that fate will repeatedly drop unpleasantness on their heads.

'Chickens Prefer Beautiful Humans'

We trained chickens to react to an average human female face but not to an average male face (or vice versa). In a subsequent test, the animals showed preferences for faces consistent with human sexual preferences (obtained from university students).

— from the published report 'Chickens Prefer Beautiful Humans'

The Official Citation

The Ig Nobel Interdisciplinary Research Prize was awarded to:

Stefano Ghirlanda, Liselotte Jansson and Magnus Enquist of Stockholm University, for their inevitable report 'Chickens Prefer Beautiful Humans'.

IG NOBEL PRIZES 2 The study was published in *Human Nature*, volume 13, number 3, 2002, pp 383–9.

Is human beauty beyond the ken of non-humans? A trio of scientists in Sweden decided to find out.

Humans, as a rule, love chickens. Some love them as pets, others as food, others as the stoic butt of jokes. The prototypical American joke, for example, consists of the question 'Why did the chicken cross the road?' followed by the answer 'To get to the other side'.

Chickens, as a rule, love humans, too. Or at least they strongly prefer some to others. Given the opportunity, chickens consistently select human beings who are lovely – certifiably lovely – over those of lesser visual charms. We know this thanks to some careful work done by Stefano Ghirlanda, Liselotte Jansson and Magnus Enquist. The research team was based at the University of Stockholm, where Ghirlanda and Jansson were graduate students working with Professor Enquist.

Is it easy to determine whether chickens prefer beautiful humans to those who are nondescript or ugly? No, not if one wishes to be confident about the answer. The task – the testing – must be done carefully. There are subtleties which must be reckoned with.

Ghirlanda, Jansson and Enquist examined the likes and dislikes of six chickens. The chickens were all of the species Gallus gallus domesticus. They were shown photographs of human faces. The humans were all of the species Homo sapiens.

The chickens were experienced computer users, well-practised in pecking at pictures offered up to them on a computer screen. However, these chickens were also, in an important sense, naïve: they had never before been asked to peck at televised faces.

Ghirlanda, Jansson and Enquist trained the chickens at recognizing some very average faces. They taught the hens to peck only at male human faces, and the cocks to peck only at female human faces. Each bird was deemed to be educated only when it was able to pass its test at least 75 per cent of the time. (On average, it took a chicken a little over seven hours of schooling to attain this high level of achievement.)

When the birds were fully matriculated, the intellectual fun began. They got to see a collection of seven human faces, some male, some female, some rather pretty, others less so. The chickens were free to peck as they chose. And they did. The hens chose handsome male faces, more often than not. The cocks chose beautiful female faces, also more often than not.

How, it may be asked, do the chickens' likings compare with those of human beings? Ghirlanda, Jansson and Enquist wanted to know that, of course. And so they recruited fourteen college students. The students were not required to peck, but simply to pick. They got to choose from among the same set of seven faces.

The students behaved, in essence, much like chickens. More often than not, they chose beauteous faces of the sex opposite to their own. Beauty, it seems, is recognized as beauty. Ghirlanda, Jansson and Enquist straightforwardly reported that 'We cannot of course be sure that chickens and humans processed the face images in exactly the same way.'

Non-specialists may wonder, 'Are these particular scientists professionally curious about anything other than chickens and beautiful humans?' The answer, a disappointment to some, is yes. They study all kinds of things. Professor Enquist, for example, is fascinated by the forces in nature that bring and keep couples together. His 1999 paper titled 'The Evolution of Female Coyness' has a nice, coyly seductive beginning:

> Females in socially monogamous species require a period of courtship before they start to reproduce. When female reproductive success depends on male assistance, such 'coy' behaviour might have evolved in response to male philandering. In this paper we use a dynamic optimisation model to demonstrate...

He has even explored how gossiping is a force that keeps couples together.

All three scientists are curious about how animals

Left to right: Magnus Enquist, Liselotte Jansson and Stefano Ghirlanda accept their Prize. Behind and above them looms a gigantic projected image of a chicken reacting to what is presumably a beautiful human. (Margaret Hart/*Annals of Improbable Research*)

behave socially amongst their own kind, and what that may say about us. In this context, the Pullets-Peck-At-Pretty-People research project arose rather naturally.

For testing the notion that chickens prefer beautiful humans, Stefano Ghirlanda, Liselotte Jansson and Magnus Enquist shared the 2003 Ig Nobel Prize in the field of Interdisciplinary Research.

The three co-authors travelled to the Ig Nobel Ceremony. Stefano Ghirlanda came from his home in Rome (he moved there after completing the beautiful work with chickens and receiving his PhD). Liselotte Jansson and Magnus Enquist travelled together from Stockholm. The journey from Stockholm, Sweden, to Cambridge, Massachusetts, has often been made by Nobel Laureates, freshly returning home from receiving their Nobel Prize from the King of Sweden. Jansson and Enquist's journey was somewhat lower-key, as well as in a contrary direction.

Speaking for the team, Stephano Ghirlanda said:

CHICKENS PREFER BEAUTIFUL HUMANS

Stefano Ghirlanda, Liselotte Jansson, and Magnus Enquist
Stockholm University

We trained chickens to react to an average human female face but not to an average male face (or vice versa). In a subsequent test, the animals showed preferences for faces consistent with human sexual preferences (obtained from university students). This suggests that human preferences arise from general properties of nervous systems, rather than from face-specific adaptations. We discuss this result in the light of current debate on the meaning of sexual signals and suggest further tests of existing hypotheses about the origin of sexual preferences.

KEY WORDS: **Facial attractiveness; Handicap principle; Receiver bias; Sexual selection**

The title page of the Prize-winning report.

> Thank you, everyone. I made a lot of false starts on this acceptance speech, but then I realized that perhaps nothing intelligent is expected. So, I will just thank our animals and everyone who supported our research, including our families; our universities; my mother who could come; my girlfriend who couldn't come; and everyone else.

Immediately after the ceremony, as the audience reluctantly meandered out of Sanders Theatre, several professors were heard to remark that each of the three co-authors was a distinctly beautiful human being. Friendly, but heated, debates ensued as to what extent the six chickens may have been distracted by the sight of Stefano Ghirlanda, Liselotte Jansson and Magnus Enquist.

And what of the haunting – and perhaps not fully solvable – question of *why* chickens prefer beautiful humans? Ghirlanda, Jansson and Enquist say that their work 'suggests that human preferences arise from general properties of nervous systems, rather than from face-specific adaptations.' Or, to put this in plain language: it's probably because bird brains are very human.

4. The Value of Imagination

Few people truly understand money. Those who do don't necessarily share quite the same understanding. Here are five money-related achievements for which Ig Nobel Prizes were awarded:

Stock Up on Junk

Profit of Doom

Tamagotchi and the World Economy

A Toothless Theory of Economics

Stock Up on Junk

In addition to being a talented, creative genius, Michael is among the most avaricious, ruthless, venal people on the face of the earth.

— a former colleague of Michael Milken, quoted in the book *The Predators' Ball*

The Official Citation

The Ig Nobel Economics Prize was awarded to:

Michael Milken, titan of Wall Street and father of the junk bond, to whom the world is indebted.

Several books describe Milken, his works, and his legacy. Two of the best are *The Predators' Ball: The Junk-bond Raiders and the Man Who Staked Them* by Connie Bruck (Simon & Schuster, 1988) and *Den of Thieves* by James Stewart (Simon & Schuster, 1991).

Before Michael Milken strode the earth, low-grade corporate bonds seemed most unlikely things to fall in love with. But Milken, an obscure man with a monotone voice and a famously bad toupee, worked magical wonders. He made the investment world fall madly, desperately, breathlessly in love with financial thingies called 'junk bonds'.

In a frenzied few years, people hurled billions and billions and billions of dollars into the riskiest, junkiest of investments. And then things went poof. Lots of people had been ruined, lots of companies had gone bust and Michael Milken had spent a little time in jail.

But the net result was that Michael Milken ended up with several billion dollars. And thus the story has a happy ending.

Michael Milken almost single-handedly created the idea that junk bonds are worth a lot of money. Even during Milken's heyday in the 1980s, few people understood or even cared what a junk bond is. What they did understand, what they did care about, was that Michael Milken said you could get rich by investing in these things.

What is a bond? A bond is a piece of paper on which a company (or a government) promises to pay a certain amount of money at a certain time. When this promise is trustworthy, the paper really is worth money. When the promise is not so trustworthy, maybe the paper is worth money, but maybe it's not; this kind of iffy promise is called a 'junk bond'.

Milken talked many small and large companies into creating these iffy promises, these junk bonds. He then talked the world into purchasing them.

Milken himself bought tons of junk bonds at low prices, then immediately resold them at much higher prices. He also made commissions when other people created, bought or sold them. In his peak year, Milken officially earned, if that's the word, $550 million. The actual total was substantially higher.

Several years after Milken's rise, the magazine *US News and World Report* admiringly described the master's golden touch:

As creator and potentate of the junk-bond market in the 1980s, Milken could scare up billions of dollars in hours, funding the takeover artists who shook up Corporate America as well as upstart companies that went on to change entire industries.

Many of the junk bonds eventually turned out to be worth little or be worthless, and people who had bought them lost their shirts and family jewels. Quite a few companies that had been backed by junk bonds collapsed.

The junk bonds in themselves were not Michael Milken's greatest accomplishment. Like all great men, he found a way to alter the world around him.

Milken was involved in, or inspired, the rapid rise and fall of many large corporations and financial institutions. Junk bonds proved to be a new and excitingly risky fuel for the game of 'hostile corporate takeovers' played by the swashbuckling high financiers of the 1980s.

Every year, Milken sponsored a lavish get-together in Beverly Hills for potential corporate raiders and junk-bond investors, plying them with food, celebrities and movie stars (Frank Sinatra, Diana Ross and the like) and, according to many reports, prostitutes. The event was known, charmingly, as The Predators' Ball.

With junk bonds and enough ruthlessness and recklessness, practically any rinky-dink tiny corporation could buy up practically any giant, stable corporation. Many tried. Some succeeded. And after one of these crazed little corporate canaries swallowed a big corporate cat, the result typically was indigestion and financial collapse.

A number of that era's large, sudden, gigantic corporate failures were made possible by junk bonds. Junk bonds also helped some of America's largest Savings and Loans associations (these are a particular type of bank) rapidly swell up to several times their natural size and then burst.

Michael Milken completely changed the economic landscape. His reward, in addition to nearly unimaginable amounts of money, was the 1991 Ig Nobel Prize in the field of Economics.

The winner could not, or would not, attend the Ig Nobel Prize Ceremony. He had a previous ten-year engagement (and an accompanying $600-million fine). His sentence for securities fraud and related offences was eventually reduced to two years of jail time, plus a $200 million fine, plus three years of probation. He also was barred for life from working in the securities industry.

After leaving prison, Milken began to act as a business consultant. Before long, the US Securities and Exchange Commission charged that some of his consultations violated the probation agreement. In 1988, Milken settled these new charges by paying the government the additional $42-million fine, plus interest on his disapproved new earnings.

Though many describe him as a topnotch swindler, at least one expert gave Michael Milken a beneficent final judgment:

> 'After a quarter century of involvement in medical and educational causes, Michael Milken is recognized as one of America's leading philanthropists. *Worth* magazine ranked him 6th

among the most-generous living philanthropists (behind Bill Gates and ahead of the Rockefeller brothers) for the three quarters of a billion dollars that he and his family have given to a wide range of causes over the past three decades.' That guilt-edged assessment is a centrepiece of Michael Milken's website (www.mikemilken.com). Presumably Milken himself wrote it.

For his ceaseless giving of lessons on the true value of junk investments, the world is truly indebted to Michael Milken.

Profit of Doom

Batra will become rich and wrong.

— University of Pennsylvania economist J. Scott Armstrong, in a 1988 review of Ravi Batra's book *The Great Depression of 1990*

The Official Citation

The Ig Nobel Economics Prize was awarded to:

Ravi Batra of Southern Methodist University, shrewd economist and best-selling author of *The Great Depression of 1990* ($17.95) and *Surviving the Great Depression of 1990* ($18.95), for selling enough copies of his books to single-handedly prevent worldwide economic collapse.

A lone man in Dallas, Texas, shocked the world. Apparently he acted alone.

His chosen weapon: two best-selling books, one published in 1985, the other in 1988, supplemented by frequent stints as a guest pundit on television and radio, and lucrative appearances on the lecture circuit.

He hoped to save the world from collapse. Apparently, he succeeded.

Many professional economists toil unnoticed, pondering questions the public might find small and abstruse. Ravi Batra is a different sort of economist.

The simplest introduction to Ravi Batra may be the one he wrote himself, and which is posted on his website (www.ravibatra.com):

> Dr Ravi Batra, a professor of economics at Southern Methodist University, Dallas, is the author of five international best sellers. He was the chairman of his department from 1977 to 1980. In October 1978, Batra was ranked third in a group of 46 'superstar economists' selected from all the American and Canadian universities by the learned journal *Economic Inquiry*. In 1990, the Italian prime minister awarded him a Medal of the Italian Senate for correctly predicting the downfall of Soviet communism.

Professor Batra has predicted many things.

In 1985, he wrote a book called *The Great Depression of 1990*, which was published by a small publishing company. The book was republished in 1987, by Simon & Schuster, a much larger, much more aggressive publishing company.

The book begins with the statement 'Few people have any first-hand idea of what a depression is like.' Professor Batra gives some disheartening details ('the stock market crashed; prices, interest rates, and wages fell like dominoes... Suddenly there was mass poverty, and soon thousands were on the verge of starvation.') He emphasizes how different, how very much worse, that is than any of the economic recessions of recent decades. Professor Batra than explains that the imminent depression will be 'the worst economic crisis in history'.

The book sold wildly, spreading like the economic panic of 1929.

In 1988, the publisher Simon & Schuster came out with a new Ravi Batra book, *Surviving the Great Depression of 1990*. The book jacket explains that:

> In his runaway national best seller *The Great Depression of 1990*, Professor Ravi Batra explained why we're headed for an unprecedented economic collapse. His new book tells you how to survive – and come out on top – while others will be lucky to keep their heads above water.

This book, too, immediately leaped onto the best-seller charts.

Professor Batra's warning did the trick. His books sold enough copies to stimulate the world economy. And so there was no great depression in 1990.

The final chapter of one book begins with these words: 'I am perhaps the only forecaster in history who fervently hopes that his prophesies turn out to be totally wrong.'

Professor Batra's fervent hopes were realized. His prophesies did turn out to be wrong.

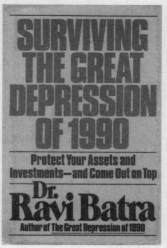

These two books by Ravi Batra apparently prevented a world financial collapse.

For this, he won the 1993 Ig Nobel Prize in the field of Economics.

The winner could not, or would not, attend the Ig Nobel Prize Ceremony.

In subsequent years, Professor Batra continued to watch the world's economic entrails, dutifully searching for dire portents.

He wrote a new series of books and articles warning of smaller-scale financial calamities. The most publicized, issued very early in 1988, was a book entitled *Stock Market Crashes of 1998 and 1999*, in which he explained how the then-recent economic woes in Southeast Asia would bring down the major stock markets elsewhere in the world. Possibly saved by

Professor Batra's warning, perhaps because his book sold enough copies to stimulate the world economy, the stock markets did not crash in 1998 or 1999.

Then, in 1999, he published a new book predicting a new momentous disaster just around the corner of the new millennium. His university issued a press release to mark the happy occasion:

> In *The Crash of the Millennium: Surviving the Coming Inflationary Depression*, Batra wrote that US financial markets would start to unravel by mid-2000 and then crash in the new millennium, with much of the damage occurring in 2000 and 2001.

The press release included a statement by the professor:

> 'I foresee a collapse of the dollar and an inflationary depression in America unfolding in the next two years. Even now perhaps this sounds like a fantastic forecast, but all my forecasts have sounded fantastic at the time they were made.'

Possibly saved by Professor Batra's warning, perhaps because his book sold enough copies to stimulate the world economy, the dollar did not collapse nor did an inflationary depression unfold in America during the next two years.

The publishing industry took note of Ravi Batra's *Crash of the Millennium* book, as it does of all likely best sellers. 'Why do so many readers listen to Professor Batra when the economy ignores him?' the trade journal *Publisher's Weekly* asked, and then immediately answered their own question: 'He is a gifted writer, which makes the dense mathematical parts of his discussion painless and the apocalyptic crescendos breathtaking.'

The newspaper *India Journal*, did a profile of the man they admiringly labelled 'Prophet of Doom'. Professor Batra posted the article on his website. The concluding words give a glimpse at the inner life of this pecuniary prophet:

> The value-based tantric meditation that Batra practises demands that he lead a saintly life on strict principles of morality with no cheating, lying, meat eating and a life dedicated to social service and opposing injustice. It also involves high-powered meditation on the cosmic entity. Unmoved by fame, the ever fertile mind is already working on his next book.

Tamagotchi and the World Economy

Your big moment is almost here – the birth of your Tamagotchi! To wake up your Tamagotchi, first you pull on the paper tab sticking out. Next, press the Reset button on the back of the egg (not too hard, now – there's a baby inside).

— from *The Official Tamagotchi Care Guide and Record Book*

The Official Citation

The Ig Nobel Economics Prize was awarded to:

Akihiro Yokoi of Wiz Company in Chiba, Japan and Aki Maita of Bandai Company in Tokyo, the father and mother of Tamagotchi, for diverting millions of person-hours of work into the husbandry of virtual pets.

A brand-new Tamagotchi in its
protective enclosure.

In November of 1996 a tiny, egg-shaped hunk of plastic
and electronics went on sale in Japan. In the next couple
of years, millions of people bought these things and
devoted large chunks of their lives to nurturing and
tending the purely digital creatures that seemed to live
inside them. Called Tamagotchi (pronounced 'ta-ma-
goh-chee' – the name means 'lovable egg'), the gizmos
dominated the days, the nights and the thought processes of countless children and a surprising number
of adults.

For a time, it almost seemed that Tamagotchis
would come to dominate all human activities. Criminal

acts were committed for love of the Tamagotchis. Schools and businesses banned Tamagotchis from their premises, fearing that the lures, charms and electronic squawks of the Tamagotchis would cause students and employees to neglect their basic, everyday activities.

The first Tamagotchi, like all its siblings and descendants, was a digital egg, but it did not come from some digital chicken. The virtual critter was conceived in the brain of a young woman and delivered by a middle-aged male product designer.

Late-twenty-something Aki Maita, who was employed in the sales and marketing bowels of the Bandai Corporation, came up with an idea that got her noticed. The inspiration, she said, was a television commercial about a little boy who insisted on taking his turtle to kindergarten. Bandai makes toys – the Mighty Morphin Power Rangers was one of their whirlwind big hits. Bandai shared Aki Maita's idea with designer Akihiro Yokoi of the Wiz Company, who whipped the idea into a lovingly detailed bundle of mass manufacturable parts.

Almost as soon as the Tamagotchi was introduced in Tokyo, it became an object of desire. Those who didn't themselves desire to play with the tiny virtual pets saw the crazed passion of those who did – and so a collectors' market sprang into existence. For quite a while, no matter how many Tamagotchis went on sale, demand far outran supply. The frenzied buying and hoarding of Tamagotchis spread almost instantly from Japan to Hong Kong and Singapore. In those three places alone, more than four million were sold during the feverish first months. Factories strained to raise the Tamagotchi birth rate.

Newspapers carried luridly cheerful details about the lengths people would go to for a Tamagotchi. Parents travelled hundreds of miles to wait in sales lines. Aggressive businessmen paid fifty times the street price. Schoolgirls mugged other schoolgirls. Movie stars made sure to have themselves photographed with their favourite Tamagotchis. A whole black market sprang up of stolen and counterfeit Tamagotchis, as did a cadre of look-almost-alike products from other manufacturers.

Like children, Tamagotchis would frequently but unpredictably 'go to sleep', 'wake up', cry for attention, sulk, defecate, and even get sick, all of this electronically and noisily. They demanded slavish attention, and they got it. People would drop whatever they were supposed to be doing, day or night, to give the things whatever they seemed to be demanding.

A few panicked authorities – and then more than a few – tried to ban them. In many schools and workplaces, officials were seriously worried by the sight of people young and old, male and female, sneakingly, then openly, dropping whatever they were doing in response to a Tamagotchi's every clarion beep for attention.

Other countries saw the news reports and prepared as best they could. It was like waiting for an epidemic of glee to arrive. A reporter for the US-based website ZDNet described what to expect, and revealed that the export-version Tamagotchis would have one significant change from the original design:

What is a Tamagotchi? It is a digital creature that

The instructions on the back of the box indicate that the Tamagotchi will need loving and elaborate care. On a global scale, this has translated into many millions of person-hours.

grows from an egg to old age. You control the progression on a small screen on an egg-shaped case. Three buttons let you manipulate the little creature, from feeding it and playing a simple game, to giving it shots of virtual medicine and scolding it. Depending on how well you take care of Tamagotchi, the virtual pet grows into one of 12 different creatures that have different personalities – from greedy to well-groomed. In the end, the creature goes back to its own planet – which is somewhere in cyberspace, according to Bandai – with a little graphic showing the creature in a

spacesuit. This is a departure from the original
version, which just out-and-out kicked the bucket.
'We didn't think kids would handle death well,' said
a Bandai spokesperson.

Bandai produced Tamagotchis in bewildering
variety, varying the colour and shape of the cases, the
behaviours of the virtual creatures and almost anything
else that might lure a consumer to buy just one more.
There grew to be more than 400 different Tamagotchi-
related products, including underwear, inner tubes and
even curry.

In the next year, Tamagotchis spread to nearly every
civilized part of the earth.

For enabling so many people to redirect so much of
their time from whatever they had previously been
doing, Akihiro Yokoi and Aki Maita, the mother and
father of the Tamagotchi, won an Ig Nobel Prize in the
field of Economics.

The winners could not, or would not, attend the 1997
Ig Nobel Prize Ceremony.

Tamagotchi sales spiked in 1997, and then, as if
toppled from atop a wall, had a great fall. Just a few
years later it was difficult to find a new Tamagotchi
anywhere, and even the collectors' market had smashed
to bits. Not all the king's horses, nor all the king's men
could have put it back together again, nor would they
even care to try.

A Toothless Theory of Economics

Robert Genco, DDS 1963, is a distinguished professor and chair of the oral biology department at UB. A renowned dental researcher, he is broadening his reach to educate customers.

— from the University of Buffalo alumni magazine

The Official Citation

The Ig Nobel Economics Prize was awarded to:

Dr Robert J. Genco of the University of Buffalo for his discovery that financial strain is a risk indicator for destructive periodontal disease.

 Dr Genco announced his discovery in a press release. Two years after winning the Prize, he published details in the report 'Models to Evaluate the Role of Stress in Periodontal Disease', Robert J. Genco, et al., *Annals of Periodontology*, volume 3, number 1, July 1998, pp 288–302. He then published further details in the report 'Relationship of Stress, Distress, and Inadequate Coping Behaviors to Periodontal Disease', Robert J. Genco, et al., *Journal of Periodontology*, volume 70, 1999, pp 711–23.

New medical insights always demand attention, and this insight was exceptional: that people with money troubles have more dental problems than people who have plenty of money.

The story drew coverage from news media around the globe. Surprising as the news was, few people were as open-mouthed as the man who had made the discovery – Robert J. Genco, DMD.

Dr Genco and his colleagues at the University of Buffalo wanted to know who has severe dental problems and who doesn't. To find out, they examined well over a thousand people – examined their teeth, their mouth bacteria and what might be called their psychosocial status.

Dentists examine teeth all the time, and some also examine bacteria when that seems called for. But how many dentists meticulously analyse their patients' psychosocial status? Dr Genco used special psychosocial questionnaires with nine questions about financial strain, including:

🏆 At the present time, are you able to afford a home that is large enough?
🏆 Do you have difficulty in meeting monthly payments of your family bills?
🏆 How often is it that you don't have enough money to afford the kind of food, clothing, medical care or leisure activities you and your family need or want?

After analysing the teeth and the bacteria and the psychosocial backgrounds, Dr Genco thought he saw a dominant fact emerge. People who were under great financial strain generally had worse dental problems than people who weren't.

But that is not quite the way Dr Genco phrased it. In a report published much later in a dental research journal, he gave a precise description, in terms that would enlighten and inspire any professional dentist:

> Reliability of subjects' responses and internal
> consistencies of all the subscales on the instruments
> used were high, with Cronbach's alpha ranging from
> 0.88 for financial strain to 0.99 for job strain,
> uplifts, and hassles. Logistic regression analysis
> indicated that, of all the daily strains investigated,
> only financial strain was significantly associated
> with greater attachment and alveolar bone loss.

Dr Genco is one of the world's most celebrated dental authorities. As another of his press releases says:

> Genco has received many awards, including the
> George W. Thorn Award from the UB Alumni
> Association (1977); William J. Gies Foundation
> Award from the American Academy of
> Periodontology (1983); Samuel P. Capen Alumni
> Award, the UB Alumni Association's most
> prestigious award (1990); American Dental
> Association Gold Medal for Excellence in Dental
> Research (1991), and the American Academy of
> Periodontology Gold Medal (1993).

And for his incisive analysis of the effects of grinding, gnawing financial distress, Dr Genco has also won the 1996 Ig Nobel Prize in the field of Economics.

The winner could not, or would not, attend the Ig Nobel Prize Ceremony.

In 1998, two years after winning the Prize, Dr Genco at last formally reported his discovery to the scientific community, publishing a detailed report in a dental research journal. Almost nobody paid attention, so the

Relationship of Stress, Distress, and Inadequate Coping Behaviors to Periodontal Disease

R.J. Genco,* A.W. Ho,* S.G. Grossi,* R.G. Dunford,* and L.A. Tedesco†

Background: The association of stress, distress, and coping behaviors with periodontal disease was assessed.
Methods: A cross-sectional study of 1,426 subjects between the ages of 25 and 74 years in Erie County, New York, was carried out to assess these relationships. Subjects were asked to complete a set of 5 psychosocial questionnaires which measure psychological traits and attitudes including discrete life events and their impact; chronic stress or daily strains; distress; coping styles and strategies; and hassles and uplifts. Clinical assessment of supragingival plaque, gingival bleeding, subgingival calculus, probing depth, clinical attachment level (CAL) and radiographic alveolar crestal height (ACH) was performed, and 8

Periodontal diseases are inflammatory conditions caused by infection with subgingival bacteria. Analysis of sys-

Several years after winning the Ig Nobel Prize, Robert Genco published this formal scientific report about his discovery.

next year he published substantially the same report in a different dental research journal.

He then issued another press release, giving it the headline:

Financial Stress Doubles Periodontal Disease Risk

In this, more clearly than in his formal published reports, Dr Genco revealed that the bad news was accompanied by a dollop of very good news.

First he gave the bad news:

Financial strain is a long-term, constant pressure. Our studies indicate that this ever-present stress and a lack of adequate coping skills could lead to altered habits, such as reduced oral hygiene or teeth grinding, as well as salivary changes and a weakening of the body's ability to fight infection.

And then the good news:

However, people who dealt with their financial strain in an active and practical way (problem-focused) rather than with avoidance techniques (emotion-focused) had no more risk of severe periodontal disease than those without money problems.

Dr Genco ended his press release, as he did each of his formal published reports, with a word of hopeful caution. 'Further studies,' he said, 'are needed.'

5. The Pursuit of Love

The sex drive and the drive to reproduce can be powerful. Sometimes it is clear that the two drives are intimately related. Sometimes it is not so clear. This chapter may shed light on the matter. Or maybe not. Here you will encounter:

Homosexual Necrophilia in the Mallard Duck

❦

Superior Sperm

❦

Clones From a Seed

❦

Homosexual Necrophilia in the Mallard Duck

Nobody had ever seen and reported this behaviour in mallards. I did, and probably that's why I'm here.

— C.W. Moeliker, in his Ig Nobel acceptance speech

C.W. Moeliker delivers his acceptance speech and displays a stuffed mallard duck. (John Bradley/*Annals of Improbable Research*)

The Official Citation

The Ig Nobel Biology Prize was awarded to:

C.W. Moeliker, of Natuurmuseum Rotterdam, the

Netherlands, for documenting the first scientifically recorded case of homosexual necrophilia in the mallard duck.

IGNOBEL PRIZES 2 The full report was published as: 'The First Case of Homosexual Necrophilia in the Mallard Anas platyrhynchos (Aves: Anatidae)' C.W. Moeliker, *Deinsea*, volume 8, 2001, pp 243–7. Truly interested parties can visit the Natuurmuseum Rotterdam and have coffee with C.W. Moeliker.

One afternoon, C.W. Moeliker heard and saw something unusual happening outside a window. Being a good scientist, he got a pen and paper, and a camera, and hunkered down to observe.

Figure 1. The new north wing of the Natuurmuseum Rotterdam. (a) marks the office of the author. (b) is the approximate location where the first duck hit the glass façade. (c) is the site from where the author observed the rape (from behind the window). (Photo: Christian Richters)

What happened is perhaps best described in dry, scientific language. The following is a brief version of C.W. Moeliker's original report. Certain abstruse technical details have been removed. The account is reproduced here with kind permission from the Natuurmuseum Rotterdam and the scientific journal *Deinsea*.

The First Case of Homosexual Necrophilia in the Mallard Duck

by C.W. Moeliker, Natuurmuseum Rotterdam, the Netherlands

The all-glass façade of the extension of the Natuurmuseum Rotterdam, situated in an urban park, acts – under certain light conditions – as a mirror (see Figure 1). Numerous birds, mostly thrushes, pigeons and woodcocks, die in collision with the building.

Especially during the first months after the new wing was erected in 1995, a 'bang' or a sharp 'tick' on the window meant work for the bird department.

The Case

Such was the case on 5 June 1995 at 17.55 hours. An unusual loud bang, one floor below my office (again, see Figure 1), indicated yet another collision and an addition to the bird collection. I went downstairs immediately to see if the window was damaged, and saw a drake mallard lying motionless on its belly in the sand, two metres outside the façade.

The unfortunate duck apparently had hit the building in full flight at a height of about three metres from the ground (yet again, see Figure 1).

Next to the obviously dead duck, another male mallard (in full adult plumage without any visible traces of moult) was present (see Figure 2a). He forcibly pecked into the back, the base of the bill and mostly

Figure 2(a) A drake mallard in full breeding plumage (right) next to the dead drake mallard now known as NMR 9997-00232, moments after NMR 9997-00232 collided with the glass-walled wing of the museum.

Figure 2(b) The same couple during copulation, two minutes after photo 2(a) was taken. (Photos: C.W. Moeliker)

into the back of the head of the dead mallard for about two minutes, then mounted the corpse and started to copulate, with great force, almost continuously pecking the side of the head (see Figure 2b).

Rather startled, I watched this scene from close quarters behind the window (see Figure 1) until 19.10 hours, during which time (75 minutes!) I made some photographs and the mallard almost continuously copulated his dead congener.

He dismounted only twice, stayed near the dead duck and pecked the neck and the side of the head before mounting again. The first break (at 18.29 hours) lasted three minutes and the second break (at 18.45 hours) lasted less than a minute.

At 19.12 hours, I disturbed this cruel scene. The necrophilic mallard only reluctantly left his 'mate': when I had approached him to about five metres, he did not fly away but simply walked off a few metres, weakly uttering a series of two-note 'raeb-raeb' calls.

I secured the dead duck and left the museum at

19.25 hours. The mallard was still present at the site, calling 'raeb-raeb' and apparently looking for his victim (who, by then, was in the freezer).

The Secured Specimen

The specimen I collected is now officially called 'NMR 9997-00232' (NMR is an abbreviation for 'Natuurmuseum Rotterdam'). Dissection revealed that the specimen is indeed of the male sex: the testes were yellow, fully developed, and measured 28x15 millimetres each.

Collision with the museum building had caused the following lethal internal damage: severe haemorrhages in the brain, rupture of the right lung, trachea and liver, both scapulae broken, most ribs broken close to the sternum (this might have been caused by the prolonged copulation). Otherwise, the duck is in good condition.

Discussion

Although I did not actually see the events preceding the moment NMR 9997-00232 hit the building and died, I strongly believe the two mallards were involved in some kind of aerial chase or pursuit flight: the victim flew into the building in full flight and the drake that pursued, managed to prevent a collision and landed next to the dead duck.

I watched the scene immediately (less than a minute) after the tremendous bang and saw the drake next to its dead congener (see Figure 2a). It is therefore highly unlikely that the drake was just passing by, saw the corpse and started to rape it.

When we disregard the homosexual nature of this case, the pursuit-behaviour the mallards were involved

C.W. Moeliker, in the vantage point from which he observed and documented the incident action. (Photo: C.W. Moeliker)

in is common (and is also often observed in the museum park). In the springtime, after the break up of the mallard pairs, when the drakes congregate in small flocks, more than a dozen may chase a single female in the air, trying to force her down and rape her.

Homosexual Rape

The biologist Bruce Bagemihl, in his well-researched and exhaustive book about animal homosexuality, showed that male homosexuality can be regarded as a common phenomenon among mallards. However, according to Bagemihl, drake pairs do not exhibit overt sexual activity: normally neither partner mounts the other.

Interestingly, Bagemihl noted that 'some males in homosexual pairs have been observed attempting to rape or forcibly copulate with' other males. Initially, this may have been the case on 5 June 1995: the drake attempted to rape NMR 9997-00232, who fled, and a chase ensued.

Necrophilia

The fact remains that NMR 9997-00232 was dead while he was being raped. (One may argue that the copulation was not rape, but certainly the act was non-consensual.) Surely, this must have influenced the duration of the copulation.

According to the scientific literature, necrophilia is known in the mallard, but only among heterosexual pairs. Occasionally, males even try to mate with dead females. To the best of my knowledge, though, this incident is the first described case of homosexual necrophilia in the mallard.

For not letting a remarkable incident pass unrecorded, C.W. Moeliker was awarded the 2003 Ig Nobel Prize in the field of Biology.

C.W. Moeliker travelled to the Ig Nobel Prize Ceremony accompanied by a stuffed duck. In accepting the Prize, he said:

Thank you. [The audience began spontaneously chanting 'Quack, quack, quack, quack!'] Quiet. Thank you very much for this, uh, award. Let me explain what I reported on. I saw two mallards. That's a duck like this [Dr Moeliker held up the stuffed mallard duck], in case you're not an ornithologist. The thing is, both were of the male sex, and one of them was dead. It had hit the façade of the Natural History Museum, Rotterdam, where I work as a curator. The living duck mounted the corpse and raped it. For more than one hour. Then, I couldn't stand it any longer. I collected the duck, dissected it, and, indeed, it was of the male sex. Nobody had ever seen and reported this behaviour in mallards. I did, and probably that's why I'm here. Thank you very much.

The following spring, he and the duck took part in

the Ig Nobel tour of the UK and Ireland. The pair of them became favourites, not only in the theatres and lecture halls of the official tour events, but also in the pubs of London, Glasgow, Dublin and several other cities.

Superior Sperm

Our Special purpose is to give babies the best possible start in life. This is how we do it: We search the nation for men in excellent health who in addition have accomplished something outstanding, or men who, though young, demonstrate great potential. Always these men have high intelligence. This, like good health, can be handed on to offspring.

— from an ad for The Repository for Germinal Choice

The Official Citation

The Ig Nobel Biology Prize was awarded to:

Robert Klark Graham, selector of seeds and prophet of propagation, for his pioneering development of The Repository for Germinal Choice, a sperm bank that accepts donations only from Nobellians and Olympians.

IGNOBEL PRIZES 2 Robert Klark Graham's book *The Future of Man*, the Christopher Publishing House, 1970, explains much of the reasoning behind his work.

Anyone who wants sperm from a Nobel Laureate has

but two good alternatives. One is to buy direct from the manufacturer. The other is – or rather, until 1999 was – to go shopping at The Repository for Germinal Choice, in Escondido, California.

Robert Klark Graham founded The Repository for Germinal Choice in 1979. He based it on the ideas of Nobel Laureate Hermann J. Muller, who had been dead since 1967, and who had argued that mankind's intelligence would degenerate unless someone did a better job of choosing who breeds with whom.

Muller won his Nobel Prize in 1946, for discovering that X-rays cause fruit-fly genes to mutate. The discovery got him thinking. Muller was not the type to keep his thoughts to himself. As described by one of his former graduate students, 'He was short (5 foot 2 inches), bald, and energetic. Some people attract trouble and thrive on controversy. Muller was one.' Muller reportedly once tried to recruit Josef Stalin as a sperm donor, but there is no record of his having succeeded.

The Repository's original name was The Hermann Muller Repository for Germinal Choice, but Muller's widow complained. Robert Klark Graham thereupon demoted Muller to a less prominent position, in which he appeared only on the organization's letterhead, listed as 'co-founder'.

The Repository advertised its wares in newsletters published by Mensa, the international group for people who score high on intelligence tests. The main pitch was rather catchy:

> We, The Repository For Germinal Choice, collect germinal material from outstandingly intelligent

and healthy men and freeze it under liquid nitrogen. The RFGC makes this germinal material available to married couples who want children but cannot have them because of the husband's infertility. Couples may choose which one of the sperm donors they prefer to become the biological father of their child. This is germinal choice.

The Repository For Germinal Choice not only enables wives of infertile men to become mothers, but increases the chances of giving their children a genetically advantaged start in life.

It also enables outstanding men to have more offspring than they would have otherwise. This puts more of our best genes into the human gene pool.

The Repository has been doing this on a pilot scale since 1980. It has helped to produce hundreds of bright and healthy children. Many have shown highly exceptional abilities.

Humankind is far from perfect but can be improved gradually by increasing the proportion of advantageous genes in the human gene pool. The Repository provides one way to accomplish this.

The current fees for recipients are the following: $100 application fee, $200 cryogenic tank fee, $3,000 program fee (for six months).

Graham maintained that three Nobel Laureates had made deposits at the bank. The only one who publicly acknowledged – in fact, bragged about – doing it was William Shockley, who won his Nobel Physics Prize in 1956 for helping develop the transistor, and who was not a shy or self-doubting man. Short on donors, Graham soon loosened the qualifications, and sent his assistant out with a small collection kit to try to procure the sperm of young, handsome male scientists who were thought to be possible *future* Nobel Prize winners. Later, still low on his basic raw material, Graham opened the doors even wider, attempting to solicit semen from accomplished artists and athletes, successful businessmen and, according to some reports, from Prince Philip of England.

The first customer to bear a Germinal Choice child was a woman named Joyce Kowalski, in 1982. Her husband told the *National Enquirer*, 'We'll begin training Victoria on computers when she's 3, and we'll teach her words and numbers before she can walk.' Press reports mentioned that Mrs and Mr Kowalski were convicted federal felons who had served jail time for a credit-card scheme based on using the identities of dead children, and who had lost custody of Mrs Kowalski's two children from a previous marriage because of alleged child abuse.

Once the Repository did get in gear, it supplied seed that grew into about fifteen or so children a year. So far as is known, none of that seed came from Nobel Laureates, and none of the progeny has yet won a Nobel Prize.

Robert Klark Graham, though, was awarded a signal honour. For his efforts to smart up humanity, he won the 1991 Ig Nobel Prize in the field of Biology.

The winner could not, or would not, attend the Ig Nobel Prize Ceremony. At the Ceremony that year, there were four Nobel Laureates on stage, handing out the Ig Nobel Prizes to the new winners. All were men; all insisted they were not depositors in The Repository for Germinal Choice.

Robert Klark Graham died in 1997. The Repository closed in 1999.

Clones From a Seed

I was a clever boy then, and I'm a clever boy now.

— Richard Seed, at his fiftieth college reunion, comparing his plans as a college student with his plan, a half-century later, to clone himself.

The Official Citation

The Ig Nobel Economics Prize was awarded to:

Richard Seed of Chicago for his efforts to stoke up the world economy by cloning himself and other human beings.

A tall, elderly man, a thundering biblical patriarch in appearance and manner, made a stunning public announcement from his modest house in Chicago.

He said he had a plan. He said he was going to clone himself.

The news media flocked to his door. They couldn't wait to hear the details. They also couldn't believe their good fortune, because the man's name was better than anything a hack novelist would have invented.

This is the real-life story of Dick Seed.

By all accounts, especially his own, Richard Seed was a brilliant child who predicted that some day he would win a Nobel Prize. He did well in school, and eventually got a PhD in physics from Harvard University.

At some point, he developed an intense and very personal interest in sexual reproduction. That led him, somehow, to develop an intense and very personal interest in asexual reproduction.

Dr Seed spent years learning how to transfer embryos

Richard Seed became an instant celebrity when he announced plans to clone himself.

from one animal to another. At first he specialized in cows. Then he applied the technique to human beings, and was involved in some of the first cases of surrogate motherhood. All this was preparation for bigger things.

On 5 December 1997, Dr Seed announced that he would clone human beings. The public's imagination had already been fired by decades of bad science-fiction movies, and more realistically by the announcement in 1996 that scientists in Scotland had actually managed to clone a sheep. Hundreds of defective cloned sheep died for every one that lived, but there was hope that perhaps the same failure rate wouldn't apply if you tried to clone people.

Perhaps, maybe, perhaps, well, yes, could be, *peut-être*, well, why think negative thoughts. Dr Seed's plan to clone people would be a giant step of some kind for mankind, should he actually carry it off. And his demeanour and tone of voice suggested that yes, Seed would succeed. And should Seed succeed, he made it clear, then Seed upon Seed upon Seed would succeed Seed.

At first, he said he had four couples who were eager to have him clone their family members. Before long, though, perhaps pricked by the many critics who scoffed at him, Dick Seed made another public announcement:

> I have decided to clone myself first to defuse the criticism that I'm taking advantage of desperate women with a procedure that's not proven.

At that time he was 69 years old. He would implant the little embryo clone into his wife, Gloria. Gloria would then give birth to an infant Richard Seed.

Gloria Seed never publicly signed onto the bandwagon.

For a time Dr Seed considered joining forces with a Canadian firm called Clonaid, which was run by a religious group called the Raelians, which worshipped a former French racing-car driver who called himself Rael, who said that extraterrestrial technologists called the Elohim resurrected Jesus using advanced proprietary cloning techniques, which were the same techniques that Clonaid would use to clone anyone who was willing to pay them a lot of money. However, the proposed partnership between Dr Seed and Clonaid didn't pan out.

Richard Seed never did clone a human being, and despite his childhood prediction, he has not yet won a Nobel Prize. But he did win the 1998 Ig Nobel Prize in the field of Economics.

Welcome to **CLONAID**tm – the first human cloning company ! **CLONAID**tm was founded in February 1997 by Raël who is the leader of the __Raelian Movement__, an international religious organization which claims that life on Earth was created scientifically through DNA and genetic engineering by a human extraterrestrial race whose name, Elohim, is found in the Hebrew Bible and was mistranslated by the word "God". The Raelian Movement also claims that Jesus was resurrected through an advanced cloning technique performed by the Elohim.

At one time, Richard Seed considered joining forces with the Raelians, a religious group that worships an extraterrestrial racing-car driver and has commercial plans to clone human beings.

The winner could not, or would not, attend the Ig Nobel Prize Ceremony, explaining that he had a previously scheduled event – a highly remunerative lecture appearance in Ireland.

Richard Seed instead sent his son Randall to accept the Prize. Randall Seed appeared bemused at, and resigned to, his place in history and in his family. He told the assembled crowd at Harvard's Sanders Theatre and the larger audience watching the event via the Internet:

> I assure you that tonight Dr Seed would be with us except that he was forced to accept that financial offer in Ireland. [At this point a heckler in the audience shouted, 'Give him a year and he'll do both at once!']
>
> Let me tell you a bit about the history. People ask, 'Is he serious?' My answer to you is that fifteen years ago, when I was living with him, my summer job was Dad handing me these four-inch Petri dishes and a microscope, and he'd say, 'Randall, go find if there's a human egg in this dish.'
>
> At any rate, I can credit him with one thing. He forced me to make a decision. I had to determine which was more disgusting, sifting through the dish

that was full of what we call uterine and vaginal fluid versus doing what we call a sperm motility test. Okay? So I decided to become an engineer.

At any rate, his latest plan is to clone himself. My brothers and I are looking forward to testing the hypothesis: is it nature or nurture? He thinks he's going to live forever by copying himself. Unfortunately, my brothers and I have things in mind to torment his life. I don't think it's ever going to be the same.

Nine months after the Ig Nobel Prize Ceremony, Richard Seed did visit Harvard, his college class's fiftieth reunion (Seed graduated from Harvard in 1949). While there, he met with a member of the Ig Nobel Board of Governors. When informed that the next Ig Nobel Prize Ceremony would feature a new mini-opera inspired by his life and philosophy, Dr Seed offered to come and perform in the opera, provided that he be paid $5000 to do so.

The 1999 Ig Nobel Prize Ceremony did include the world premiere of *The Seedy Opera*, which was all about Richard Seed and his legend. Dr Seed was not in the cast, having priced himself out of the local opera-singer market. In the opera, the fictional Richard Seed made multiple clones of himself – but first he polished up his technique by cloning some sheep. The opera began with him singing the following words, to the tune of 'O Sole Mio':

I made some new sheep
    Didn't need a ewe-sheep
    In the world of science
    I have joined the giants.

Were I a glutton

In an opera based loosely on Richard Seed's life, the great man's mother (sung here by mezzo-soprano Margot Button) sings of her ever-increasing pride in her self-reproducing brood. (Photo: John Chase, Harvard News Office)

> I could feast on mutton
> But it's not worth it for a few bytes of RAM.

O solo me-oh! Oh, lonesome me!
> Although a genius I well may be,
> For a genius
> It's ignomeenious
> To have these stupid sheep for company.

The Seedy Opera concluded with legions of Richard Seeds forming a giant corporation to manufacture and sell even more copies of themselves. The opera ended with them singing to the apocalyptic tune from *Carmina Burana*:

Here's what we'll do –
We'll sell them to
 Armies of every nation.
No need to draft,
With armies staffed
 By clones of our creation.

When there's a war
We'll just make more.
 How many will you nee-eed?
We will provide
To either side
 As many copies as they want of Richard Seed.

(*Lyrics*: Don Kater and Marc Abrahams)

6. Medical Mysteries Solved

When one sheds light on bodily secrets, one sometimes finds the unexpected. Here are five who shed light, and then found themselves honoured with an Ig Nobel Prize.

Scrotal Asymmetry in Man and in Ancient Sculpture

The Possible Pain Experienced During Execution by Different Methods

Impact of Wet Underwear on Comfort in the Cold

Nicotine is Not Addictive

Preference for Waxed or Unwaxed Dental Floss

Scrotal Asymmetry in Man and in Ancient Sculpture

Winckelmann in 1764 commented that: 'Even the private parts have their appropriate beauty. The left testicle is always the larger, as it is in nature.' He went on, however, 'so likewise it has been observed that the sight of the left eye is keener than the right,' an observation which, to my knowledge, has not been confirmed.

To test Winckelmann's claim, I observed the scrotal asymmetry of 107 sculptures, either of antique origin or Renaissance copies, in a number of Italian museums and galleries...

— From the published report by Chris McManus

The Official Citation

The Ig Nobel Medicine Prize was awarded to:

Chris McManus of University College London, for his excruciatingly balanced report, 'Scrotal Asymmetry in Man and in Ancient Sculpture'.

IGNOBEL PRIZES 2 His report, entitled 'Scrotal Asymmetry in Man and in Ancient Sculpture,' was published in *Nature*, volume 259, 1976, p. 426.

The sculptors of ancient Greece, famed to this day for their care and skill, got something terribly and – now that we know about it, thanks to Chris McManus – embarrassingly wrong.

When Chris McManus was a young doctor, he took a

summer off to travel from his native England and wander the hills and cities, and especially the museums and art galleries, of Italy. He scrutinized, but did not physically handle, a total of 107 sculptures of men. Some of the pieces were of ancient origin, others were copies made during the Renaissance. He noted, on a page of his notebook, two things about each statue: (a) which testicle was larger and (b) which was the higher.

At the end of the summer, he looked over his notes, and realized that the conventional wisdom about such things (expressed in the phrase 'The left testicle is always the larger, as it is in nature') was wrong. As any learned medical person might do, he decided to correct the record.

Scrotal asymmetry in man and in ancient sculpture

MITTWOCH and Kirk[1] have claimed that "Right and left mammalian gonads do not usually differ noticeably either in

*Present address and address for reprin Gardens, Kenton, Harrow, Middlesex

[1] Mittwoch, U., and Kirk, D., *Natur* (1975).
[2] Chang, K. S. F., *et al.*, *J. Anat.*, 94,
[3] Winckelmann, J. J., in *History* (transl. by Gode, A.)(Frederick U₁ 1968).
[4] Lloyd, G. E. R., *J. Hellenic Stud.*, 8

Table 1 Analysis of the scrotal asymmetry of 107 ancient sculptures

		Left	Side of higher testicle Equal	Right
Side of larger testicle	Left	2	7	32
	Equal	8	19	17
	Right	17	1	4
	Total	27	27	53

Chris McManus's Prize-winning paper.

Dr McManus wrote a brief account of it all, and submitted his report to the journal *Nature*. *Nature* is, of course, one of the – some say *the* – most prestigious science journals in the world, and among the most difficult in which to get a paper accepted for publication. Many scientists would kill to get a paper published in *Nature*. Some would kill repeatedly for the chance.

Nature accepted the article for publication, and then surprised Dr McManus by featuring it on their cover.

Here is Dr McManus's fond look back, almost three decades later, at how his scrotal adventure unfolded:

Half a lifetime ago, in 1976, I published in *Nature* a curious little paper entitled 'Scrotal Asymmetry in Man and in Ancient Sculpture'. Despite being only 353 words long, and hidden deep in the bottom-right-hand corner of a page at the back of the magazine, somehow it avoided the total obscurity which should have been its fate.

The essential result was replicated a few months later by Andrew Stewart, a genuine classical scholar, then at the start of his career, who had studied *kouroi*, the much earlier set of pre-classical sculptures, and who has subsequently gone on to be the doyenne of scholars of classical sculpture. The *Nature* paper also provoked a discussion of whether Rodin could be diagnosed as dyslexic on the basis of the conspicuously reversed scrotum in his sculpture *The Age of Bronze* (a reversal not visible, it must be said, in photographs of the well-endowed Belgian soldier, Auguste Neyt, who was the model).

Attention then waned, and I did little with the topic itself, other than including a more detailed chapter in my PhD thesis, although I have continued to research the topic of lateralisation. A few aficionados sometimes mentioned the scrotal asymmetry paper during the coffee breaks of conferences, and there was the occasional scholarly reference in books on lateralisation, or, particularly

Chris McManus accepts the Medicine Prize, honouring his insights into scrotal asymmetry in man and ancient sculpture. (Photo: Ruby Arguilla/ Harvard News Office)

pleasingly, on Greek sculpture.

I thought that my most notorious paper had gone gently into the good night of oblivion until in October 2002 it was resurrected for an Ig Nobel Prize for Medicine, stimulating subeditors the world over to spawn bad headlines, my favourite being 'Oddball scientist wins prize'.

For taking a good, round look at what's left and what's right, Chris McManus was awarded the 2002 Ig Nobel Prize in the field of Medicine.

Dr McManus travelled to the Ig Nobel Ceremony. In accepting the Prize, he read part of a poem he had composed for the occasion. Here is a manly extract from that literary work. An understandably rare marriage of medicine and literature, in form and pacing it is based on Henry Wadsworth Longfellow's tortuous classic, *Hiawatha*. Perhaps schoolchildren will recite this poem in literature and science classes, and on formal celebratory occasions.

Ball Park Estimate
by Chris McManus

By the thigh tops, by the groin ends,

Just in front of the perineum,
Lies the sac they call the scrotum,
With the vasa deferentia,
And the epididymes.

In that scrotum lie the testes
On the right side and the left side,
At the first glance seem symmetric
Left and right side each the same size.

That may seem to be straightforward,
But it isn't, have a good look,
Close inspection shows the diff'rence:
Use the eyeball, check the right ball,
Clearly higher than the left ball;
Also size balls with the eyeballs,
Though the right ball may be higher
It is also bigger, larger.
Without doubt the balls are diff'rent,
Symmetry is not the case here.
Ancient Greece – they liked their young boys,
Loved them even, called them *kouroi*.
Even Socrates and Plato,
Dined with them at their symposia.

In museums find their statues
White and smooth of Paros marble,
Sculpted in their every detail
Every muscle, every sinew,
Front and rear view, limbs and torso.

Got them right but made one error:
Got the balls wrong! – Even though the
Right was higher (as it should be),
There was still a chiral error,
Left the bigger – not the smaller.

Ideas were the basic problem,
Ideas dating back for aeons,
Talked about by all Greek thinkers,
Hippocratic or Platonic,
Even by the Pre-Socratics,
For Pythagoras it seemed so,
And Anaxagoras also,
Even to Parmenides, and
To Empedocles as well, that
Right side differs from the left side.

Aristotle gave the reasons.
In synopsis, what he said was
That the balls had but one function,
Act as weights and stretch the body,
Pull the larynx, make the voice drop,
Introducing adolescence.

As a theory it was quite good
But it had an implication.
If the left ball's lower, heavier
It must also be the big ball
So the right must be the *smaller*.
Though the sculptors thought it made sense
There's an error, that's apparent.
To describe it, Greek is no use,
Here one cannot say *'Eureka!'*
But instead needs Anglo-Saxon:
It's an error, it's a cockup!
Some would say they 'made a balls-up'.

What went wrong here, what's the problem,
Where's the error that misled them?
Observations disregarded,
Ideas triumphed over data,

> Theories may be nice and pretty
> But there's more than that to science.
>
> At the end we need a moral
> Something true while yet Ig-Nobel:
> Don't be misled by a model
> (Be it theory or those *kouroi*...).

Several months after receiving the Ig Nobel Prize, Dr McManus was awarded another honour – the Aventis Prize, which is meant to 'celebrate the very best in popular science writing for adults and children'. Stephen Hawking, Jared Diamond, Stephen Jay Gould and Roger Penrose are among the previous winners. Dr McManus won for his book *Right Hand, Left Hand: the Origins of Asymmetry in Brains, Bodies, Atoms and Cultures*, which is based partly on the insights he had gained from researching 'Scrotal Asymmetry in Man and in Ancient Sculpture'.

His curiosity over little things had become, decades later, a matter for high public celebration. From small nuts do mighty oaks grow.

A museum visitor contemplates the asymmetric properties of a statue in the Victoria and Albert Museum in London. (Photo: Stephen Drew/*Annals of Improbable Research*)

The Possible Pain Experienced During Execution by Different Methods

The physiology and pathology of different methods of capital punishment are described. Information about this physiology and pathology can be derived from observations on the condemned persons, postmortem examinations, physiological studies on animals undergoing similar procedures, and the literature on emergency medicine. It is difficult to know how much pain the person being executed feels or for how long, because many of the signs of pain are obscured by the procedure or by physical restraints...

— from the published report by Harold Hillman

The Official Citation

The Ig Nobel Peace Prize was awarded to:

Harold Hillman of the University of Surrey, England, for his lovingly rendered, and ultimately peaceful, report 'The Possible Pain Experienced During Execution by Different Methods'.

IG NOBEL PRIZES 2 His report was published in *Perception* 1993, volume 22, pp 745–53.

There exist few reliable, first-hand reports of the pain experienced during an execution. Thanks to Harold Hillman, we now at last have an extensive – if not quite first-hand – report of the *possible* pain experi-

The possible pain experienced during execution by different methods

Harold Hillman
Unity Laboratory of Applied Neurobiology, University of Surrey, Guildford GU2 5XH, Surrey, UK
Received 21 January 1992, in revised form 29 June 1992

Abstract. The physiology and pathology of different methods of capital punishment are described. Information about this physiology and pathology can be derived from observations on the condemned persons, postmortem examinations, physiological studies on animals undergoing similar procedures, and the literature on emergency medicine. It is difficult to know how much pain the person being executed feels or for how long, because many of the signs of pain are obscured by the procedure or by physical restraints, but one can identify those steps which

The Prize-winning report.

enced during execution by each of the most popular methods.

Harold Hillman was Director of the Unity Laboratory of Applied Neurobiology, and a Reader in Physiology at the University of Surrey, in Guildford. He spent years gathering information to demonstrate the pain and suffering inherent in several forms of capital punishment.

Dr Hillman drew from a wide variety of sources: 'observations on the condemned persons, postmortem examinations, physiological studies on animals undergoing similar procedures and the literature on emergency medicine'.

This he caringly distilled into a fact-filled, eight-page report that provoked reactions of many different kinds – admiring, disgusted, disdainful, horrified, and in some circles, mordantly *amusé*.

Dr Hillman gave a detailed description of each method of execution – how the act is performed, the typical physiological course of events in the executee and a quick pathological examination of the remains.

He began with shooting. ('This may be carried out [by an executioner] who fires from behind the condemned person's occiput towards the frontal region...')

Next came hanging.

After hanging came stoning. Dr Hillman pointed out that 'This form of execution is likely to result in the slowest form of death of any of the methods used.'

Stoning was followed with beheading. ('The skin, muscles, and vertebrae of the neck are tough, so that beheading does not always result from a single blow...')

Immediately after beheading came electrocution.

Then came gassing. ('The condemned person is strapped to a chair in front of a pail of sulphuric acid, in an airtight chamber...')

And in the end, came intravenous injection. ('The condemned person is bound supine to a trolley and a trained nurse or technician cannulates the vein in the angle of the elbow...')

Having described, in quite gory detail, the nuts and bolts of each form of execution, Dr Hillman then got to the heart of the matter: the pain. Dr Hillman makes no wild claim to omniscience. As he put it: '[One does not] know for how long and how severely a decapitated head feels. There are substantial areas of ignorance, so that one cannot know for certain the extent of pain in respect of a particular method.'

What one can do, Dr Hillman pointed out, is watch for *signs* of pain. He got specific: 'In everyday life, a person in severe pain shouts or screams, perspires, has dilated pupils, withdraws from the noxious stimulus, moves the limbs violently, contracts the facial muscles, micturates, and defecates.'

Dr Hillman constructed a helpful little chart to show, at a glance, which of these signs of possible pain can typically be detected during which methods of execution. (See chart on page 120.)

Altogether, the Hillman report presents a useful treasury of grisly detail, augmented with medical speculation. Its ultimate conclusion: 'All of the methods for executing people, with the possible exception of intravenous injection, are likely to cause pain.'

The report, together with the massive research involved in producing it, earned Harold Hillman the 1997 Ig Nobel Peace Prize.

Citing a combination of ambivalence and financial constraints, the winner chose not to attend the Ig Nobel Prize Ceremony. Six years later, though, he took part in the Ig Nobel Tour of the UK and Ireland, where he delighted and mystified audiences in several cities.

iGNOBEL PRIZES 2

The Possible Pain for Each Method

Here is a reproduction of Dr Hillman's chart indicating which form of pain can be observed in which method of execution.

(+) indicates that the sign of distress is often seen,

(–) indicates that the sign is seldom or never seen, (*) indicates that the sign of distress is sometimes seen but sometimes not, and (?) indicates that Professor Hillman's research did not determine whether this sign of distress is typically seen.

	Shooting	Hanging	Stoning	Beheading	Electrocution	Gassing	Injection
Shouting or screams	?	*	+	+	*	+	–
Perspiration	?	?	*	?	+	?	?
Dilated pupils	*	?	*	?	*	?	+
Withdrawal from stimulus	*	*	*	*	*	*	*
Violent movements	?	+	*	*	*	*	–
Contraction of facial muscles	*	+	*	?	*	+–	–
Micturation	?	+	?	?	+	?	?
Defecation	?	+	?	?	+	?	?

The Impact of Wet Underwear on Comfort in the Cold

The purpose of this study was to investigate the significance of wet underwear...

— from Bakkevig and Nielsen's published report

The Official Citation

The Ig Nobel Public Health Prize was awarded to:

Martha Kold Bakkevig of Sintef Unimed in Trondheim, Norway, and Ruth Nielsen of the Technical University of Denmark, for their exhaustive study, 'Impact of Wet Underwear on Thermoregulatory Responses and Thermal Comfort in the Cold'.

IGNOBEL PRIZES 2 Their report was published in *Ergonomics*, volume 37, number 8, August 1994, pp 1375–89.

Wet underwear, when worn in cold weather, has an impact on thermal comfort. Martha Kold Bakkevig and Ruth Nielsen performed the first good, scientific analysis of this chilling phenomenon.

Bakkevig and Nielsen were intent on understanding underwear, with the long-range goal of improving it. They performed their research methodically.

First they got eight men who were willing to wear wet underwear in the cold while having their skin

Impact of wet underwear on thermoregulatory responses and thermal comfort in the cold

MARTHA KOLD BAKKEVIG

SINTEF UNIMED, Section for Extreme Work Environment,
N-7034 Trondheim, Norway

and RUTH NIELSEN

Laboratory of Heating and Air Conditioning,
Technical University of Denmark, DK-2800 Lyngby, Denmark

Keywords: Thermoregulatory responses; Subjective sensation;
Rest; Cold; Underwear; Sweating.

The purpose of this study was to investigate the significance of wet underwear and to compare any influence of fibre-type material and textile construction of underwear on thermoregulatory responses and thermal comfort of humans during rest in the cold. Long-legged/long-sleeved underwear manufactured from 100% polypropylene in a 1-by-1 rib knit structure was tested dry and wet as part of a

The Prize-winning report.

and rectal temperatures monitored. Each man had a surface area of approximately two square metres.

To avoid influencing their behaviour, Bakkevig and Nielsen did not tell the men any details about what kinds of underwear they would be wearing or how cold it would get.

And it did get chilly. The experiment was carried out in a special test chamber where the temperature was kept at 10 degrees centigrade (50 degrees Fahrenheit). Some men were given a specially prepared set of wet underwear. Others were given dry underwear. The underwear was of various kinds – wool, cotton, polypropylene and various blends.

The night before, the researchers had put the underwear and some water into an airtight bag, which they then sealed and kept in a warm oven overnight.

This produced what is called 'a satisfactory distribution of water' in the underwear.

Before donning the soggy undergarments, each man was weighed in the nude, and temperature sensors were affixed to his forearm, neck, chest, abdomen, upper back, lumbar back, upper arm, forearm, anterior thigh, shin, calf, the dorsal side of one hand, and the dorsal side of one foot. A rectal thermometer was inserted 80 millimetres to a depth beyond the anal sphincter. The rectal thermometer was 150 millimetres long, a model YSI 701 manufactured by Yellow Spring Instruments, of Dayton, Ohio.

Each man sat alone in a chair in the cold room for 60 minutes with the thermometer up his butt. Every ten minutes he filled out a questionnaire which asked:

Are you:
    1. heavily shivering?
    2. moderately shivering?
    3. slightly shivering?
    4. not at all shivering/sweating?
    5. slightly sweating?
    6. moderately sweating?
    7. heavily sweating?

How do you feel thermally:
    1. comfortable?
    2. slightly comfortable?
    3. uncomfortable?
    4. very uncomfortable?

There were additional questions, all along similar lines.

Every 60 seconds, machines recorded the man's

skin temperature, weight and rectal temperature. At the end of the hour, the man removed the wet underwear and was again weighed in the nude. The underwear was also weighed.

The scientists analysed all the data, and produced graphs and charts depicting what had happened, statistically, during the course of a typical hour spent sitting alone wearing wet underwear in a cold room.

Some things remained constant throughout the course of the hour:

🍸 Men wearing wet underwear always reported that their underwear felt wet.

🍸 Men wearing dry underwear always reported that their underwear felt dry.

🍸 Men with wet underwear felt colder than men with dry underwear. They also felt less comfortable.

Bakkevig and Nielsen drew two conclusions from their research:

First, that wet underwear *does* influence thermo-regulatory responses and thermal comfort in the cold.

And, *second*, perhaps more surprisingly, that underwear's thickness matters much more than what it's made of.

For discerning a truth about wet underwear, Martha Kold Bakkevig and Ruth Nielsen won the 1995 Ig Nobel Prize in the field of Public Health.

The winners could not, or would not, attend the Ig Nobel Prize Ceremony. Eventually, though, they did express modest, if ambivalent, delight at being honoured.

Nicotine is Not Addictive

I believe nicotine is not addictive.

— testimony in the US *Congressional Record*, 15 April 1994

The Official Citation

The Ig Nobel Medicine Prize was awarded to:

James Johnston of R.J. Reynolds, Joseph Taddeo of US Tobacco, Andrew Tisch of Lorillard, William Campbell of Philip Morris, Edward A. Horrigan of Liggett Group, Donald S. Johnston of American Tobacco Company and the late Thomas E. Sandefur, Jr, chairman of Brown and Williamson Tobacco Co. for their unshakable discovery, as testified to the US Congress, that nicotine is not addictive.

IG NOBEL PRIZES 2 Their testimony was published in the US *Congressional Record* for 15 April 1994.

After years of intense prodding from the American public, medical professionals, and a succession of US Surgeons General, the United States Congress held public hearings to investigate the health effects of cigarette smoking.

First, a parade of medical and scientific experts presented evidence that cigarette smoking is probably not the most healthful of activities.

Then on 14 and 15 April 1994, the chief executive officers of the seven major tobacco companies were brought to the Capitol building in Washington, DC.

A panel of congressmen asked the seven CEOs to explain how their companies manufacture and market cigarettes.

The hearings centred on a handful of questions, most pointedly:

1.   whether cigarettes cause health problems;
2.   whether the companies build their marketing activities to take advantage of the habit-forming properties that cigarettes are known to have; and
3.   whether the companies manipulate nicotine levels in their cigarettes to try to make them more habit-forming.

At first, the seven CEOs spoke with masterful, thoroughly professional obfuscation. But eventually one congressman asked a simple question that drew a uniform, simple answer from all of them.

Here is a transcript of their testimony to the US House Energy and Commerce Committee Subcommittee on Health and the Environment.

REPRESENTATIVE RON WYDEN: Let me ask you first, and I'd like to just go down the row, whether each of you believes that nicotine is not addictive. I heard virtually all of you touch on it. Just yes or no. Do you believe nicotine is not addictive?

MR CAMPBELL: I believe nicotine is not addictive, yes.

REP. WYDEN: Mr Johnston?

MR JAMES JOHNSTON: Congressman, cigarettes and nicotine clearly do not meet the classic definitions of addiction. There is no intoxication.

REP. WYDEN: We'll take that as a no and, again, time is short. If you can just – I think each of you believe nicotine is not addictive. We just would like to have this for the record.

MR TADDEO: I don't believe that nicotine or our products are addictive.

MR HORRIGAN: I believe nicotine is not addictive.

MR TISCH: I believe that nicotine is not addictive.

MR SANDEFUR: I believe that nicotine is not addictive.

MR DONALD JOHNSTON: And I, too, believe that nicotine is not addictive.

For this testimony, the seven CEOs were awarded the 1996 Ig Nobel Prize in Medicine.

The winners could not, or would not, attend the Ig Nobel Prize Ceremony.

Preference for Waxed or Unwaxed Dental Floss

The purpose of this study was to discover patient preference for waxed or unwaxed dental floss, and to learn more about individual flossing habits. One hundred patients randomly presenting for routine dental examinations volunteered to sample a brand of similar-appearing waxed and unwaxed dental floss. After flossing an anterior and a posterior contact area with both types, the patients indicated whether they preferred the waxed or unwaxed floss.

— from Beaumont's published report

The Official Citation
The Ig Nobel Dentistry Prize was awarded to:

Robert H. Beaumont, of Shoreview, Minnesota, for his

incisive study 'Patient Preference for Waxed or Unwaxed Dental Floss.'

IGNOBEL PRIZES 2   His study was published in the *Journal of Periodontology*, volume 61, number 2, February 1990, pp 123–5.

Which do people prefer – dental floss that is waxed or dental floss that is unwaxed?

There are several ways to approach this question. Manufacturers of dental floss might ask which of them people actually purchase. Public health officials might ask which is the more effective at keeping teeth and gums healthy.

Dr Robert H. Beaumont, the Chief of Periodontics at the 842nd Strategic Hospital, at Grand Forks Air Force Base in North Dakota, tackled a third aspect of the question. Dr Beaumont stripped the inquiry to its rawest, most basic, form. Purchasing habits and health effects aside, he wondered which kind of dental floss do people simply, in their hearts, *prefer* – waxed or unwaxed?

Even the briefest account of dental-floss research must begin with a grateful mention of Charles C. Bass: 'Charles C. Bass was responsible for the early development of specifically formulated unwaxed nylon floss and the popularization of personal oral hygiene techniques in common use today.' Those words were written by Dr Beaumont, more than 35 years after Dr Bass intrigued the dental world with his powerful vision of filamentous nylon.

(The full story of dental floss has yet to be written.

Patient Preference for Waxed or Unwaxed Dental Floss*

Robert H. Beaumont

THE PURPOSE OF THIS STUDY was to discover patient preference for waxed or unwaxed dental floss, and to learn more about individual flossing habits. One hundred patients randomly presenting for routine dental examinations volunteered to sample a brand of similar-appearing waxed and unwaxed dental floss. After flossing an anterior and a posterior contact area with both types, the patients indicated whether they preferred the waxed or unwaxed floss. The patients also answered questions concerning their flossing

The Prize-winning report.

Robert H. Beaumont, author of the scientific report 'Patient Preference for Waxed or Unwaxed Dental Floss'.

It involves more than one intriguing personality, and several unexpected turns. Space limitations do not permit a full account here.)

Dr Beaumont carried out his dental floss preference research on 100 patients.

First, he gave each patient a length of dental floss dispensed from an unmarked container, with instructions to floss between two front teeth, then to floss between two back teeth.

Then he gave them a length of visually similar floss dispensed from a different, also unmarked, container, and had them floss in the same places.

Dr Beaumont scientifically randomized the procedure. Half the patients were given first waxed floss, then unwaxed. The other patients got unwaxed, then waxed.

Finally, Dr Beaumont asked each patient to say which of the two pieces of floss he or she liked better. As his report later summed up the results: 'All had an immediate and clear floss preference after performing the test.'

Seventy-nine per cent of those patients preferred

waxed floss. Twenty-one per cent preferred unwaxed.

Afterwards, Dr Beaumont debriefed each of the patients, asking for their qualitative assessments of the floss. He recorded this information and included a summary in his published report. For anyone with teeth, the following passage is worth chewing over:

> The most frequent objection to waxed floss was related to a feeling of 'thickness', not a specific objection of difficulty in use. Unwaxed was described as 'thinner' and when preferred was most often selected for that reason.

For his straightforward approach to dental floss preference research, Dr Robert H. Beaumont was awarded the 1995 Ig Nobel Prize in the field of Dentistry.

Dr Beaumont could not attend the Ig Nobel Prize Ceremony, but he did send a recorded speech. In accepting the prize, he said:

> You know, an Ig Nobel Prize is certainly better than a No Nobel Prize. The world now knows that 79% of the patients in my study preferred waxed dental floss over the unwaxed variety. Unfortunately it was also discovered that only 29.5% of the patients claimed to floss daily. An astounding 17% do not even floss once a week. This is a sad statistic. I hope that my paper can serve as a stimulus to improve those numbers. I finally wish to thank my wife whose idea was the inspiration for this study, and without whose support this would have been impossible. Thank you, I think.

In a conversation with the Ig Nobel Board of Governors, Dr Beaumont divulged the main reason he had carried out his research project. He said that at the time he was in the military and had 'a lot of free time' on his hands.

IGNOBEL PRIZES 2 **Which is Better?** Dr Beaumont's report examined only whether patients prefer waxed or unwaxed dental floss. He did not carry out experiments as to which is more effective. He believed that the large body of previous research had already settled that question. As Dr Beaumont put it in his report: 'An unsubstantiated belief in the superiority of unwaxed floss has persisted to the present.'

7. Peace and Quiet

Peace and quiet do exist, now and then, in one place or another. Here are three Ig Nobel Prize-winning achievements related to the pursuit of one or the other or both:

Have a Pepsi, Win a Prize, Start a Riot

Good Nukes Make Good Neighbours

The Zipper-entrapped Penis

Have a Pepsi, Win a Prize, Start a Riot

Even if I die here, my ghost will come to fight Pepsi. It is their mistake. Not our mistake. And now they won't pay. That's why we are fighting.

— A protester in Manila, explaining to the *Los Angeles Times* why she continued to participate in anti-Pepsi protests despite her own ill health and after her husband had died of heart failure following an anti-Pepsi protest rally.

The Official Citation

The Ig Nobel Peace Prize was awarded to:

The Pepsi-Cola Company of the Philippines, suppliers of sugary hopes and dreams, for sponsoring a contest to create a millionaire, and then announcing the wrong winning number, thereby inciting and uniting 800,000 riotously expectant winners, and bringing many warring factions together for the first time in their nation's history.

IG NOBEL PRIZES 2 One source of information about the Pepsi contest is a website maintained by a group of disgruntled ticket-holders. The group is called the Coalition for Consumer Protection and Welfare. Their Pepsi 349 website is at www.iconex.net/pepsi349.

Pepsi and Coke, Coke and Pepsi. The two great cola companies waged a continual war against each other in the Philippines, as they did everywhere else. In early 1992, Pepsi launched a major new offensive, a contest called Number Fever. Gambling contests are popular in the Philippines, and so this seemed like a good idea. However.

For Pepsi and for its customers, Number Fever started out as a lot of fun. It was simple, and potentially very lucrative. Every bottle cap had a number printed in the inside. About one out of every 50,000 bottles had some kind of winning number and the prizes ranged as high as a million pesos (equivalent to about US$40,000). By buying a single bottle of Pepsi, you would have about a one-in-28,000,000 chance of winning a one-million-peso prize. The more bottles you bought, the greater your chance – albeit still extremely, extremely slim – of becoming a millionaire. The advertising slogan was, in fact, 'Today, You Can Be a Millionaire!' Altogether there were ten of the top prizes and a host of much smaller ones.

The contest was a big, big hit. Pepsi sales surged. Coke sales plummeted.

The head of Pepsi-Cola Products Philippines, Inc. (PCPPI) waxed ecstatic, telling reporters that Number Fever was 'the most successful marketing promotion in the world. Half of the country's population was involved in it. There is no other promotion in the world that attracted that high a participation rate.' Customers were buying as many bottles of Pepsi as they could afford, many buying a case of 24 bottles every day for weeks on end. For Pepsi, this was lucratively exciting.

Seizing the moment, the company decided to extend the contest. It would select a few new winning numbers, and give away eight new one-million-peso prizes.

But something went a little wrong.

On 25 May 1992, the company announced that one of the new winning numbers was 349, and as usual the evening television news spread the word to the entire country.

The problem, as was quite soon brought to their attention, was that 349 had been a *losing* number in the first phase of Number Fever, and about 800,000 people had bought bottle caps imprinted with the number 349. Each of those 800,000 bottle caps was now worth one million pesos.

What happened next can be described at great length, as it has been in court documents, police reports, and Philippine newspaper accounts. It can also be summarized concisely, although at the sacrifice of some interesting and colourful details, by saying: riots broke out.

Here is a slightly longer, but still compressed version.

Crowds gather at the nation's twelve Pepsi bottling plants. Angry crowds. Pepsi-Cola Products Philippines, Inc.'s board meets through the night, trying to decide what to do. The board calculates that even if all of the bottle caps are redeemed, the company will have to pay out 800,000,000,000 pesos (US$32 billion), an amount larger than the combined valuation of all the companies of all kinds listed on the Manila stock exchange. The board decides to disqualify 349 as the winning number, and pick a new winning number. The board announces this. The crowds are not pleased. There are riots. At least two people die. At least 38 Pepsi trucks are damaged. Reports say that many Pepsi company officials have fled the country. The company offers to pay each holder of a 349 bottle cap 500 pesos rather than one million pesos; the offer has a time limit of two weeks. Several million cap holders accept the 500 pesos. Many millions do not. Lawsuits are filed. Criminal charges are filed. The combined number of civil and criminal cases rises above the ten-thousand mark.

For uniting so many people in a single cause, the Pepsi-Cola Company of the Philippines won the 1993 Ig Nobel Peace Prize.

The winners could not, or would not, attend the Ig Nobel Prize Ceremony.

Many of the legal proceedings against the company were resolved over the course of the next few years, but more than a decade later, many of the legal wheels continue to grind, and spin and spin.

Good Nukes Make Good Neighbours

The world community should appreciate the fact that India, the second most-populous country on earth, waited for five decades before taking this step.

— Indian Prime Minister Shri Atal Bihari Vajpayee, quoted in the *New York Times*, 16 May 1998

Today we have evened the score with India.

— Pakistani Prime Minister Nawaz Sharif, quoted in the *New York Times*, thirteen days later

The Official Citation

The Ig Nobel Peace Prize was awarded to:

Prime Minister Shri Atal Bihari Vajpayee of India and Prime Minister Nawaz Sharif of Pakistan, for their aggressively peaceful explosions of atomic bombs.

Alphonse and Gaston are traditional comic characters, each politely insisting the other take the first step. Although the characters are traditionally French, the English equivalent, 'Keeping up with the Joneses' is a pleasant, hackneyed phrase, a slightly colourful way of describing neighbours who love to outdo each other. The expression is thought to be American. A fine real-life example of this behaviour, occurred in southern Asia, in India and Pakistan, in the years leading up to 1998.

Alphonse and Gaston, in the guise of India and Pakistan, were endlessly inviting each other – on this, that, and the other thing, and always purely out of politeness – to go first.

Before the taking of a first step, each was operatically insistent that the other go first, be it to fire the first shot, launch the first mortar, send in the first wave of troops, drop the first bomb, buy the first jet fighter, buy the newer-model jet fighter, or fire the first long-range missile.

On each go-round, after weeks, months, or years, either Alphonse reluctantly got on with it, or Gaston did.

In 1998, the newest new step was to explode a nuclear bomb. Neither had exploded one, and each politely insisted that the other would have to do it first. The level of politeness was quite wonderful – each insisted that only the other even had a bomb. (Actually, India had exploded a show-off bomb years before, but, by 1998, the Alphonse-Gaston ritual ignored that by mutual consent.)

As always happens with this particular Alphonse and this particular Gaston, one of them did, finally, take the step.

On 11 May 1998, India set off three underground nuclear explosions. Completing the switch from Alphonse mode to Impress-the-Joneses mode, two days later they set off another pair of bombs.

It took Pakistan seventeen days to switch from Gaston mode to Keep-Right-Up-There-With-Those-Joneses Mode. On 28 May, they set off *five* atomic bombs. Pakistan's Prime Minister Nawaz Sharif said: 'Today we have settled the score with India.' Two days later, he remembered that Keeping Up With the Joneses usually involves a ritual Go-the-Joneses-One-Better-ing. And so on 30 May, Pakistan exploded one more atomic bomb.

Each set of Joneses now had fulfilled its Jonesly responsibilities. The little spending spree was over, and Alphonse went back to being Alphonse, Gaston to being Gaston.

For their bit in keeping up with the folks over the hedge, Shri Atal Bihari Vajpayee and Nawaz Sharif shared the 1998 Ig Nobel Peace Prize.

The winners could not, or would not, attend the Ig Nobel Prize Ceremony.

In a curious, loose intertwining of Ig Nobel destinies, six years later, Prime Minister Vajpayee fought off an electoral challenge in his home district from a member of the Association of Dead People. This occurred a mere eight months after the Association's founder, Lal Bihari, was awarded the Ig Nobel Peace Prize – and it was at Lal Bihari's urging that his legally-dead compatriot entered the electoral arena. In that congressional election, Prime Minister Vajpayee did manage to keep his political career alive. His challenger remained, technically as well as politically, dead.

The *Zipper-entrapped* Penis

The views expressed in this article are those of the authors and do not reflect the official policy or position of the Department of the Navy, Department of Defense, nor the US Government.

— from the published report by Nolan, Stillwell, and Sands

The Official Citation

The Ig Nobel Medicine Prize was awarded to:

James F. Nolan, Thomas J. Stillwell, and John P. Sands, Jr, medical men of mercy, for their painstaking research report, 'Acute Management of the Zipper-entrapped Penis'.

IGNOBEL PRIZES 2   Their study was published in the *Journal of Emergency Medicine*, volume 8, number 3, May/June 1990, pages 305–7.

ACUTE MANAGEMENT OF THE ZIPPER-ENTRAPPED PENIS

James F. Nolan, MD, Thomas J. Stillwell, MD, and John P. Sands, Jr., MD

Departments of Urology and Clinical Investigation, Naval Hospital, San Diego, California
Reprint address: LT. J. F. Nolan, MC, USNR, c/o Clinical Investigation Department, Naval Hospital, San Diego, CA 92134-5000

☐ Abstract — A zipper-entrapped penis is a painful predicament that can be made worse by overzealous intervention. Described is a simple, basic approach to release, that is the least traumatic to both patient and provider.

☐ Keywords — zipper; foreskin/penile skin; bone cutter

INTRODUCTION

Uncircumcised young boys occasionally catch their foreskin in the process of zipping or unzipping clothing. A simple method for extraction of the male foreskin entrapped in a zipper, which is presented in this case report, has been noted previously in the pediatric and

briefly anesthetized for the zipper removal. Using the jaws of a strong bone cutter, the median bar of the zipper fastener was cut and the upper and lower shields of the device separated, releasing the skin with minimal resultant injury (Figure 2). A formal circumcision was then undertaken at the parents' request.

DISCUSSION

The foreskin of the uncircumcised male, and less often the redundant penile skin of the circumcised male, may be entangled by a zipper. This occurs most commonly in the downward unzipping movement, but can also occur, as in our patient, with the upward closing of the zipper. Males who go without protective underclothing and

Nolan et al.'s Prize-winning report.

In just thirty-three words, Drs James F. Nolan, Thomas J. Stillwell and John P. Sands, Jr summarized the work that would bring them recognition throughout the medical profession:

> A zipper-entrapped penis is a painful predicament that can be made worse by overzealous intervention. Described is a simple, basic approach to release that is the least traumatic to both patient and provider.

It all began with a simple incident, that of a young boy whose ventral prepuce was lodged in the zipper of his sleeper pyjamas. The doctors used the jaws of a strong bone-cutter device to cut the median bar of the zipper fastener, thus freeing that which was trapped. 'A formal circumcision,' they report, 'was then undertaken at the parents' request.'

The doctors published a formal report so that other physicians, if confronted by similar cases, would not be caught unawares.

Here is their explanation of what can happen:

The foreskin of the uncircumcised male, and less often the redundant penile skin of the circumcised male, may be entangled by a zipper. This occurs most commonly in the downward unzipping movement, but can also occur, as in our patient, with the upward closing of the zipper. Males who go without protective underclothing and children wearing night clothing with anterior zippers are at a higher risk for this injury.

The three doctors imply that training and experience are of value in treating this condition. 'Extraction by vigorous manipulation,' they write, 'including attempts at unzipping the skin or prying the zipper are usually unsuccessful, painful, and can lead to further injury. Our case exemplifies a simple, quick, and nearly pain-free method of freeing entrapped zippers.'

For their skilful handling of a delicate danger, James F. Nolan, Thomas J. Stillwell and John P. Sands, Jr, won the 1993 Ig Nobel Prize in the field of Medicine.

Dr James Nolan travelled from his office in western Pennsylvania to the Ig Nobel Prize Ceremony. In accepting the Prize, he said:

I wish my mother was here to see me accept this prize. My colleagues and I never dreamed this simple paper would attract so much attention. I was here to save my generation from penile injury. Your recognition here tonight has stimulated my interest in further pursuing research in the field of painful penile predicaments. My colleagues and I at the Navy Hospital in San Diego – where we performed the research – and a competing group at the

University of California, San Francisco, have already shed more light on the management of the human bite to the penis. And now, as I change my career path to be a urologist in rural America, my new colleagues and I hope to further clarify the incidence and significance of urologic trauma secondary to farm animals.

Dr Nolan has since moved to Fayetteville, North Carolina, where his presence is a great comfort to the male members of the community.

8. Voilà

Scientific discovery is, sometimes, triumphant. Here are four triumphs in which scientists brought new understanding where previously there had been only water, cannon-fired chickens, squashed bugs, or a yearning for sudden riches:

The Shower-curtain Mystery

Chicken Plucking and Tornado Wind Speed

That Bug on Your Windshield

Base Elements Into Gold

The Shower-curtain Mystery

Until now, explanations for the shower curtain's movements were theoretical. It was one person's opinion versus another's, with most ideas drawing on the Bernoulli effect or on so-called buoyancy effects. The Bernoulli effect is the principle that explains how an airplane's wings produce lift. It says that as a fluid accelerates, the pressure drops. But the Bernoulli effect is based on a balance between pressure forces and acceleration, and does not allow for the presence of droplets. Nor, according to my calculations, is it responsible for the curtain deflection.

— David Schmidt, quoted in *Scientific American*, 2001

The Official Citation

The Ig Nobel Physics Prize was awarded to:

David Schmidt of the University of Massachusetts for his partial solution to the question of why shower curtains billow inwards.

IGNOBEL PRIZES 2 Technical details of David Schmidt's shower-curtain research are available from the University of Massachusetts Department of Mechanical and Industrial Engineering.

David Schmidt did not get annoyed when a shower curtain billowed inwards. Shower curtains do that, and David Schmidt did what any good engineer would do. He analysed the hydrodynamics. This was not a cursory act. This was to be the most comprehensive analysis that anyone, ever, had done of an inwards-billowing shower curtain.

Professor David Schmidt specializes in the physics of sprays, cavitation, and other so-called 'multiphase flows'. Much of his work concerns what happens when, say, fuel is injected into the combustion chamber of an engine. His training and experience paid off when he tackled the problem of why his mother-in-law's shower curtain always wrapped itself round him.

The general question had been considered by scientists of earlier generations. There were two main camps of thought, one opining that the hot air inside the shower rises, causing cold air outside the curtain to flow inwards in a sort of chimney effect. The competing theory drew on the famous Bernoulli Principle that fast-flowing air inside the shower, compared with slow-moving air outside, induced a pressure differential across the curtain.

Professor Schmidt pondered this as the water beat upon him and the curtain drew near. He would use his expertise. He would use his computer. He would, in short, take a modern stab at the question.

Here, in his own words, is how he did it:

To do the calculation, I drafted a model of a typical shower and divided the shower area into 50,000 minuscule cells. The tub, the shower head, the curtain rod and the room outside of the shower were all included. I ran the modified software for two weeks on my home computer in the evening and on weekends (when my wife wasn't using the computer). The simulation revealed 30 seconds of actual shower time. When the simulation was

A schematic view of Professor Schmidt's shower-curtain analysis. Diagram courtesy of Fluent, Inc., developers of the software package Schmidt used to make his break-through.

complete, it showed that the spray drove a vortex. The centre of this vortex – much like the centre of a cyclone – is a low-pressure region. This low-pressure region is what pulls the shower curtain in.

He discovered that the driving spray created a vortex – a whirlwind, a miniature tornado confined to the corner of a bathroom. This vortex, he discovered, seems to be the dominant force on the shower curtain. Previous generations of scientists were not wrong; Professor Schmidt believes that they simply didn't realize that their theories covered only part of the question.

The Schmidt analysis took physics to a place it had seldom gone so wetly, examining degrees of detail unattainable to earlier generations of engineers and physicists. Yet, his may not be the last word on the matter. Shower heads come in many types, styles and efficiencies, and enclosures can have varied shapes and sizes. These are all factors to be considered. And some day, more powerful computers will allow some future David Schmidt to divide a shower volume into 500,000 minuscule sections, or 5,000,000, or some still higher multiple, rather than the 50,000 with which he made his breakthrough.

For shedding light and massive amounts of computational power onto the swirling mist, David Schmidt won the 1999 Ig Nobel Prize in the field of Physics.

The winner travelled to the Ig Nobel Prize Ceremony at his own expense. In accepting the Prize he put on a plastic shower cap, and said:

> Thank you. The Ig Nobel Prize is the greatest award I could have won for this. I do computational fluid dynamics for my living, but I wanted to take advantage of that and do something that was purely to satisfy intellectual curiosity. The fact that there was no grant involved, no contract, no deliverables, made it all the more fun. It made the work entirely mine.

Chicken Plucking and Tornado Wind Speed

One way of estimating the wind in a tornado vortex is to determine by experiment what air speed is required to blow all the feathers off a chicken, a phenomenon known to occur in severe storms.

— from Bernard Vonnegut's Prize-winning report

The Official Citation

The Ig Nobel Meteorology Prize was awarded to:

Bernard Vonnegut of the State University of Albany, for his revealing report, 'Chicken Plucking as Measure of Tornado Wind Speed'.

IGNOBEL PRIZES 2   His study was published in
Weatherwise, October 1975,
p 217.

In considering what can happen to a dead chicken when
it is fired from a cannon, Bernard Vonnegut used a
potent tool: his common sense.

The conventional wisdom about chicken-plucking as a
way to measure tornado wind speed can be found in
H.A. Hazen's book *The Tornado*, published in 1890.
Hazen gives the following account of an experiment
conducted in 1842 by Elias Loomis, a professor at
Western Reserve College in Ohio:

> The stripping of fowls attracted much attention in
> this and other tornadoes. In order to determine the

Chicken Plucking as Measure of Tornado Wind Speed

B. VONNEGUT, *Atmospheric Sciences Research Center, State University of New York at Albany, Albany, New York 12222*

...ne way of estimating the wind in a tor-... vortex is to determine by experiment ...t air speed is required to blow all the ...ers off a chicken, a phenomenon (Flora, ...; Ludlum, 1970) known to occur in these ...re storms. Hazen (1890) gives the fol-...ing account of such an experiment carried ...by Elias Loomis in 1842.

...The stripping of fowls attracted much ...attention in this and other tornadoes. ...In order to determine the velocity needed ...o strip these feathers, the above six-...pounder was loaded with five ounces of

termine what is known about the forces required to remove feathers. It was found that the force with which the feathers are held by the follicles is highly variable and in the circumstances of a tornado might be greatly reduced. It depends not only on the bird's health and molting period, but also the state of its nervous system (Voitkevich, 1966). A response known as "flight-molt" is recognized in which during conditions of stress the bird's follicles relax so that the feathers can be pulled out with far less force than is normally required (Payne, 1972). Possibly this may be a mechanism for sur-

Bernard Vonnegut's Prize-winning report.

Elias Loomis, who performed the original chicken-plucking experiment in 1842. 133 years later, Bernard Vonnegut determined that Loomis made a significant error.

velocity needed to strip these feathers, the above six-pounder was loaded with five ounces of powder, and for a ball a chicken just killed. Professor Loomis says, 'The gun was pointed vertically upwards and fired. The feathers rose twenty or thirty feet, and were scattered by the wind. On examination, they were found to be pulled out clean, the skin seldom adhering to them. The body was torn to small fragments, only a part of which could be found. The velocity was 341 mph. A fowl, then, forced through the air with this velocity is torn entirely to pieces; with a less velocity, it is probable most of the feathers might be pulled out without mutilating the body.'

This was the state of the art before other means were devised to measure the speed of a violent wind.

More than a century after Loomis's famous experiment, Bernard Vonnegut, a physicist at the State University of New York at Albany, analysed it in depth.

Vonnegut was by that time a famous man himself in meteorological circles. He and a colleague had invented what is still the best method for seeding clouds to cause rain. Bernard Vonnegut was also, among other things, an inspiration to his younger brother Kurt, a novelist of some repute.

Bernard Vonnegut saw two difficulties with the Loomis method of firing chicken carcasses from cannons.

First, he wrote, 'Because it is difficult to separate the effects produced by the explosion in the gun from those that are the result of the movement of the bird relative to the air, this version of the experiment leaves much to be desired.'

Second, he pointed out, a bird's feathers are some-times much easier to pluck than others, depending on the bird's health and moulting period, but also on 'the state of its nervous system'. There is, Vonnegut wrote, 'a response known as "flight-moult",' in which during conditions of stress the bird's follicles relax so that feathers can be pulled out with far less force than is normally required. Possibly this may be a mechanism for survival, leaving a predator with only a mouth full of feathers and permitting the bird to escape.

'In light of the fact that the force required to remove the feathers from the follicles varies over a wide range in a complicated and unpredictable way and depends on the chicken's condition and his reaction to his envir-onment, the plucking phenomenon is of doubtful value as an index.'

Vonnegut concluded from this that 'Probably it is not indicative of winds as intense as might at first be supposed.'

For decisively overturning one of science's oldest

unchallenged assumptions, Bernard Vonnegut won the 1997 Ig Nobel Prize in the field of Meteorology.

The winner passed away several months before the Ig Nobel Ceremony, but his five sons and a delegation of friends and colleagues travelled to the Ig Nobel Prize Ceremony, and accepted the Prize in his honour. His son Peter spoke on behalf of them all:

> As you probably can understand, some of the family are just plain ashamed of all of this. None of us took after our father. We are very far from science. I don't know if he would have been disgusted or delighted by this, but a bunch of us are here, and we really love having this kind
> of fun at his expense. Thank you.

The Vonnegut brothers then attempted a demonstration on stage of how to properly fire a chicken carcass from a cannon aimed straight upward. The demonstration was halted prior to detonation by the Ig Nobel Board of Governors, who are ever mindful that the Cambridge Fire Department, which has its headquarters next door to Sanders Theatre, discourages the firing of cannon inside large old brick-and-wood theatres.

That Bug on Your Windshield

Some species are actually attracted to highways (e.g., lovebugs), and others are lured in by car headlights (e.g., moths). Consequently, many insects are unintentionally hit by motorists... Fortunately (or unfortunately, depending on how one looks at it), this event provides a unique opportunity to identify and learn about these wonderful insects.

— from the book *That Gunk on Your Car*

The Official Citation

The Ig Nobel Entomology Prize was awarded to:

Mark Hostetler of the University of Florida, for his scholarly book, *That Gunk on Your Car*, which identifies the insect splats that appear on automobile windows.

IGNOBEL PRIZES 2   This book is published by Ten Speed Press, 1997.

Mark Hostetler wanted to discover what bugs were where, and in what profusion. To accomplish this task, he studied the end product of the interaction of bug and automobile.

As a graduate student in entomology (the study of insects), Hostetler haunted bus stations across North America. When riding a bus, he would look not through the windshield but directly at it. At each bus stop, Hostetler would carefully collect the bounty that had accumulated on the vehicle's forward expanse of window, grille, and headlights.

For a scientist interested in bugs, the front of a bus can serve as a massively efficient collection instrument. Its large surface area, conveyed at high speed through a bug-rich atmosphere, exhibits an economy of scale that puts to shame the nets, bags and jars used by small-scale collectors.

All things being double-edged, the front of a bus does have a disadvantage. It yields up fragments of bug, or puree of bug, rather than bug *intacto*. For Mark Hostetler, this was not a severe problem. He taught himself to identify the various kinds of splotches. Mosquitoes, midges, blowflies, grasshoppers, cockroaches, dragonflies – all have their own distinctive signatures.

In sudden death, a bug has no time for a melodious swan song; at best it can manage a short, percussive coda. It is poignant that after an incessantly musical lifetime, the creature leaves an artistic legacy that is purely visual. It paints, with impressive rapidity, a stippled canvas beautiful mostly to the educated observer.

Hostetler accrued a considerable knowledge of bug splats, then distilled it all into a book 126 pages long. The book's glory, the reason it is treasured by collectors, or will be some day, at least by collectors of bug splats, is to be found smack near the centre, sandwiched firmly after page 54 and before page 55. There Hostetler has put twenty-seven magnificent colour plates.

As there are almost uncountably large variations even within a particular species, Hostetler devised a compromise method for representing each prototypical bug splat:

As you can imagine, I had a difficult time allocating one particular splat to a particular species. More often than not, I identified a splat to a group of insects (e.g. mosquitoes or cockroaches), rather than to a known species. When trying to identify a splat, keep in mind that a splat can be quite variable; however, each splat normally has several characteristics that are specific to a particular insect group, and I list the most prominent aspects of each particular splat.

Here are three of the many:

ANTS: Usually a watery, small, white smear about 8 to 15 millimetres in length.

BUTTERFLIES AND MOTHS: Butterflies and moths usually leave a thick, gooey, white or yellow substance with lumps in it. The splat is usually strung out from the point of impact (10 to 90 millimetres), and one can see scales (dust-like particles) scattered around the permimeter.

LACEWINGS: Lacewings will always leave a long, thin, greenish line (up to 10 centimetres in length) with a small blob at one end.

For his careful analysis of bug splats, Mark Hostetler won the 1997 Ig Nobel Prize in the field of Entomology.

The winner travelled at his own expense from Gainesville, Florida, to the Ig Nobel Prize Ceremony. His fiancé, Meryl Klein, sat in the audience, and afterwards reported that she had sat next to an elderly couple who, she said, obviously made a regular practice of coming to the Ig Nobel Ceremony. The couple had brought with them a stack of paper to make into aeroplanes. Klein said that there seemed something peculiar about the paper, and at some point she finally worked up the nerve to examine it. The couple had gone to the trouble of collecting old tax forms. That was the moment, Klein said, when she truly understood the Igs.

Base Elements Into Gold

It goes without saying that my work couldn't be bogus or I wouldn't be doing it.

— John Bockris, interviewed by *The Bryan-College Station Eagle*, December 1993

The Official Citation

The Ig Nobel Physics Prize was awarded to:

John Bockris of Texas A&M University, for his wide-ranging achievements in cold fusion, in the transmutation of base elements into gold, and in the electrochemical incineration of domestic rubbish.

IGNOBEL PRIZES 2 For some of his cold fusion work see 'Tritium and Helium Production in Palladium Electrodes and the Fugacity of Deuterium Therein', J.O'M. Bockris, et al., Proceedings of the Third International Conference on Cold Fusion, 21–5 October 1992, Nagoya Japan, in *Frontiers of Cold Fusion*, H. Ikegami, editor, volume 231, 1993. Also see 'On an Electrode Producing Massive Quantities of Tritium and Helium', C-C Chien, D. Hodko, Z. Minevski and J.O'M. Bockris, *Journal of Electroanalytic Chemistry*, volume 338, 1992, p 189.
For some of his transmutation work, see 'Anomalous Reactions During Arcing Between Carbon Rods in Water', R. Sundaresan and J.O'M. Bockris, *Fusion Technology*, volume 26, 1994, p 261.

Some of John Bockris's peers say he is brilliant, tenacious, and arrogant. These are not necessarily bad traits if one wants to make great discoveries, especially if one is trying to turn base metals into gold.

John Bockris was for many years an imperious and respected professor of chemistry at Texas A&M University. Late in his career, Professor Bockris began working on some spectacular kinds of science.

First, Professor Bockris announced that he had achieved cold fusion – using a small agglomeration of test tubes to produce nuclear fusion at room temperature.

Here's the lowdown on cold fusion, for those who remember the name and the fuss, but not the details. Most physicists thought nuclear fusion was hard to pull off – and very expensive, requiring billions of dollars to create temperatures as hot as the inside of the sun. Then, in 1989, Stanley Pons and Martin Fleishmann at the University of Utah unveiled a stunning idea – a simple, cheap way to do it. They called their method 'cold fusion'. It turned out that Pons and Fleishmann had been sloppy and hadn't actually gotten the thing to work. Around the world, hundreds of other scientists tried too – very carefully – but nearly all of them failed, and most concluded that cold fusion was just a cute idea that happened to be wrong.

John Bockris, on the other hand, said that he himself was exceedingly careful, and that, in his lab, cold fusion worked just fine, no sweat. Other scientists at his university, and a parade of still others from all over, came to Professor Bockris's lab to take a look. Nearly all of them said they saw nothing. Professor Bockris was incensed, and began working at a furious pace and issuing a stream of announcements about his continuing success.

And then he took into his lab and under his wing a mysterious researcher. The mystery man was working to turn base metals into gold. He would transform iron into gold. He would transform mercury into gold. He had an even more mysterious financial backer who would pay Professor Bockris a lot of money to help.

Alchemy has been a dream for centuries, but (with the possible exception of some chickens, pigs, and lobsters – for details, see the work of the 1993 Ig Nobel Physics Prize winner Louis Kervran) apparently no-one had ever successfully pulled it off. With the rise of science over the last three centuries, people had actually been testing out claims rather than accepting the sacred and sonorous word of magicians – and, in that time, few, if any, reputable scientists have reported seeing gold appear alchemically.

The mysterious researcher in Professor Bockris's lab turned out to be a colourful character named Joe Champion, who later wrote a book called *20th Century Alchemy*, doing the bulk of the writing after he took up residence in the Maricopa County Jail, in Phoenix, Arizona.

The even more mysterious financial backer turned out to be named William Telander, who was being investigated by the US Securities and Exchange Commission for fraud in connection with an international finance scam.

Various little setbacks, as well as annoying public criticism from many of his fellow chemists, hindered Professor Bockris in his efforts to achieve both cold fusion and the turning of base metals into gold.

For his persistence in the face of scientific

discouragement, John Bockris won the 1997 Ig Nobel Prize in the field of Physics.

The winner could not, or would not, attend the Ig Nobel Prize Ceremony.

A few years later, Professor Bockris very reluctantly retired from Texas A&M University, at the strong urging of many of his colleagues.

9. Aroma Everywhere

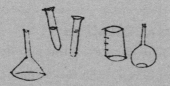

Smell is believed to be the most primitive of our senses. This chapter describes four Ig Nobel-winning attempts at aromatic sophistication:

DNA Cologne

The Perfume on Magazine Pages

Beano

Foot Malodour

DNA Cologne

This product does not contain deoxyribonucleic acid ('DNA'). Ce produit ne contient pas d'acide désoxyribonucléique ('AND').

— warning printed on the packaging for each bottle of DNA Cologne

The Official Citation

The Ig Nobel Chemistry Prize was awarded to:

Bijan Pakzad of Beverly Hills, for creating DNA Cologne and DNA Perfume, neither of which contains deoxyribonucleic acid, and both of which come in a triple-helix bottle.

DNA Cologne is available in stores, and directly from Bijan Fragrances, Inc., of Beverly Hills, New York and London. Professor Jon Marks's analytical study of DNA Cologne was published as 'Arrivederci, Aroma: An Analysis of the New DNA Cologne', which is included in the book *The Best of Annals of Improbable Research*, Marc Abrahams, editor, W.H. Freeman and Co., 1997.

In 1952, James Watson and Francis Crick discovered that each molecule of deoxyribonucleic acid – the chemical more catchily known as 'DNA' – is made of simple pieces strung together in a long double-helix arrangement. This was a stupendous discovery, because it let scientists understand how genes work – how information is passed chemically from parents to child, from each generation to the next.

Ten years later, Watson and Crick were awarded a Nobel Prize.

More than thirty years after that, DNA inspired something even more momentous, or at least more fragrant: the perfumer Bijan of Beverly Hills created DNA Cologne.

Bijan's DNA Cologne is expensive and smells, to most people, rather nice. It comes in an attractive glass helix bottle. Where the famous biological DNA molecule consists of two strands twisted round each other – a double helix – Bijan's DNA Cologne bottle goes one better. The bottle has the shape of *three* strands twisted round each other – a triple helix.

The packaging assures the consumer that DNA Cologne does not, in fact, contain any deoxyribonucleic acid. Jon Marks of Yale University analysed DNA Cologne, and concluded that it also does not smell like the biological entity after which it is named. Marks pointed out that this is in the tradition of the trade. The perfume named Poison, by Christian Dior, does not smell like poison; nor does the cologne named Stetson, by Coty, smell like a hat.

Nor, for that matter, can DNA Cologne be easily obtained at the University of Cologne, even at their Institute of Genetics, where scientists labour day and night to understand the subtle and beautiful workings of the DNA molecule.

DNA Cologne can be obtained from the manufacturer or from any authorized sales agent thereof. It makes a lovely gift for any unscented or improperly-scented scientist.

For transforming deoxyribonucleic acid into a fragrant concept, Bijan Pakzad won the 1995 Ig Nobel Prize in the field of Chemistry.

Bijan Pakzad had a previously scheduled rendezvous with a glamorous actress, and so could not come to the Ig Nobel Prize Ceremony. In his place, Sally Yeh, the president of Bijan Fragrances, Inc., travelled from company headquarters on Rodeo Drive, Beverly Hills, to Harvard, to take part in the Ig Nobel Prize Ceremony. She handed out samples of DNA Cologne to the audience, and then accepted the Prize on behalf of Bijan Pakzad. She said:

> It is my honour and pleasure to accept this award on behalf of the world-renowned, most exclusive menswear and fragrance designer, Bijan. Now, Bijan is unable to attend, but he is very excited about this event. Right now he is filming his new advertising campaign with the famous actress Bo Derek [who had become famous from her role in the movie *10*]. Together, they should make a perfect twenty. Undeniably humbled by the presence of such notable personalities in this room, I have only this to say about DNA. In our mind, DNA not only stands for 'deoxyribonucleic acid' – it also captures the initials of Bijan's three children, Daniela, Nicolas and Alexandra. They are the only ones who don't

have to go to department stores to get designer jeans.

Actually, when Bijan created these award-winning fragrances he had in mind for DNA to be interpreted as 'Definitely Not Average' and 'Damn Near Affordable'. So on behalf of the designer Bijan and everyone at Bijan Fragrances, I thank you for this award.

The Ig Nobel Board of Governors also arranged a special tribute from Nobel Laureate James Watson. Watson was aware of DNA Cologne's existence, and had once telephoned Bijan Fragrances asking for a sample, but the company didn't know who he was, and refused to send him any perfume. The Ig Nobel Board of Governors telephoned Bijan Fragrances, explained who Watson was, and arranged for them to send him some DNA Cologne.

Watson prepared an audiotaped personal tribute to DNA Cologne, to be played at the Ig Nobel Ceremony. In it, he alludes to his and Francis Crick's famous discovery, which they made at Cambridge University and announced in an article in *Nature* magazine. Here is his tribute:

When I got to Cambridge in 1951, Francis Crick told me one of the essential tricks of how he did science. He said if an idea was good, it smelled right. When we got the double helix, it smelled right. What I have to ask now is: would the double helix have received better reception if we had sprayed DNA perfume on the manuscript we sent off to *Nature*?

IG NOBEL PRIZES 2 — DNA in Song!

The 1995 Ig Nobel Prize Ceremony featured many tributes to DNA, including this song:

DEOXYRIBONUCLEIC ACID
(Lyrics by Don Kater. Sung to the tune of Richard M. and Robert B. Sherman's tune Supercalifragilisticexpialidocious.)

Backups: Some sing: Um diddle D diddle N diddle A

Others sing: Um D N A

Solo: There's an acid known as deoxyribonucleic
Known around the world from Vladivostok to Passaic.
What made quite a lot of biochemistry archaic?
It's the acid known as deoxyribonucleic!!

Backups: Some sing: Um diddle D diddle N diddle A

Others sing: Um D N A

Guanine, thymine, cytosine, (*modulate up here*)
Plus a lot of adenine

Solo: Hail an acid known as deoxyribonucleic
Known in every language from Chinese to Aramaic.
You can call it DNA, but that is so prosaic.
We prefer to call it deoxyribonucleic!!

All: We prefer to call it deoxyribonucleic!!

The Perfume on Magazine Pages

Incorporate Arcade interactive sampling technologies into your advertising and promotional vehicles and see: reading time will increase 10-fold as your target customer 'gets engaged' in looking, touching, smelling, trying and even wearing your product. The possibilities are practically endless!

Fragrance? Flowers? Food? Fun? Our scent and aroma sampling systems – from the Arcade-pioneered ScentStrip® Sampler, to ScentSeal®, DiscCover®, MicroDot™ and LiquaTouch® – deliver a superior aroma rendition of any product where the scent helps to generate the sale. The fun part? Our MicroFragrance® technology lets kids, teens and grown-ups, too, 'scratch 'n' sniff' to get a whiff of cookies, cereal, candy; nearly any food or beverage product.

— from an ad for Arcade, Inc., of Chattanooga, Tennessee

The Official Citation

The Ig Nobel Chemistry Prize was awarded to:

James Campbell and Gaines Campbell of Lookout Mountain, Tennessee, dedicated deliverers of fragrance, for inventing scent strips, the odious method by which perfume is applied to magazine pages.

IGNOBEL PRIZES 2 A partial history of scent strips can be obtained from Ron Versic, president of the Ronald T. Dodge Company of Dayton, Ohio.

Scent strips, the smellable ads in magazines, seem to

annoy more people than they please. This is not entirely
surprising; the earliest scent strips were intentionally
made to smell foul and repulsive.

Scent strips became a workhorse of the perfume-
marketing industry almost immediately after the very
first perfume scent strip magazine ad appeared in 1983,
for Giorgio perfume. Giorgio was touted as being a
glamorous, dramatic, floral fragrance sold in distinctive,
bold yellow- and white-striped packaging in tribute to
the striped awnings of the glamorous, dramatic, floral
Giorgio boutique on Rodeo Drive in Beverly Hills. But
really, the perfume was a hit because of the scent-strip
ads.

This transformed both the perfume industry and the
fashion-magazine industry, and made life hell for thousands of mail delivery people, who would often come
home smelling like a dozen different brands of perfume
had gotten on them, which it had.

Though they cause much annoyance, scent strips
also cause much demand for expensive perfume, and so
they remain a presence in perfume marketing budgets
and, therefore, on the pages of magazines.

The perfume is, in essence, printed onto the page,
the fragrance having first been microencapsulated.
Microencapsulation is a chemical process that turns a
containerful of liquid into a containerful of millions
and billions of microscopic spheres, each containing a
tiny little amount of the original liquid.

There is very big money in this technology of small
things. Because of that, and because the demand for
microencapsulating perfume rose so high so quickly in
the wake of Giorgio, there were lawsuits galore as to
who owned the rights to various parts of the process.

Because there were so many lawsuits, and so many eventual legal settlements in which the participants signed agreements binding them to not discuss details, the history of microencapsulation and scent strips is obscured in a perfumey mist. But the mist is not sufficiently dense to hide the big stink that started it all.

In the 1950s, natural-gas companies began adding a foul-smelling chemical called mercaptan to their gas, so that leaks would be very noticeable. Mercaptan is a sulphur compound with an odour reminiscent of rotten eggs, onions, garlic, skunks, and/or halitosis. To educate the public that this particular stink means *'Gas leak!'* someone came up with the idea of putting a small amount of mercaptan on a card, with instructions to scratch and sniff it. This was the first Scratch-n-Sniff card, and it was printed by a commercial printing company in Tennessee. Years later, James and Gaines Campbell, the sons of the man who owned that printing company in the 1950s, advanced the state of the art, creating the mixed blessing that is scent strips. The company, now called Arcade, Inc., is responsible, one way or another, for a good many of the scent strips that befoul modern post offices, magazine racks, and waiting rooms.

For creating a new way for literature to stink, James Campbell and Gaines Campbell won the 1993 Ig Nobel Prize in the field of Chemistry.

The winners could not, or would not, attend the Ig Nobel Prize Ceremony, but a former colleague, Ron Versic, came and accepted the Prize on their behalf. He also explained the Campbells' reluctance to come themselves. Mr Versic is now president of a company which specializes in microencapsulation. James and Gaines Campbell, he said, were parties to various restrictive legal agreements, and for that reason could not discuss the history of scent strips in any but the most general and vague terms. Mr Versic pointed out that he himself was not a party to those agreements, and consequently does not suffer the same restrictions.

Mr Versic, by the way, was not a popular man at the Ig Nobel Ceremony, because he brought with him a giant atomizer with which he insisted on spraying perfume at everyone. The perfume was not of a pleasing fragrance.

IG NOBEL PRIZES 2 What to Do With Scent Strips

Moments after the Ig Nobel Prize was awarded to James and Gaines Campbell for inventing scent strips, Nobel Laureate William Lipscomb, who is a chemist, rose to make a personal announcement:

'The international science community is worried about the future of the earth's environment. The uncontrolled release of perfume into the atmosphere constitutes a major biohazard. Please – after you have read a magazine, don't throw your smelly scent strips in the trash. Recycle them. If your town doesn't recycle scent strips, please do the next best thing – stick your scent strips in an envelope and mail them back to the magazine. Thank you.'

Beano

A few drops on your favourite (but gas-producing) food almost always stops the gas before it starts! Ends the discomfort and social embarrassment which come from eating BEANS, CHILI, CAULI-FLOWER, CHICKPEAS, SOY FOODS, and many others. Unlike other anti-gas products BEANO liquid prevents the bloat, gassiness, discomfort and embarrassment BEFORE THEY START!

— Information printed on a package of Beano

The Official Citation

The Ig Nobel Medicine Prize was awarded to:

Alan Kligerman, deviser of digestive deliverance, vanquisher of vapour, and inventor of Beano, for his pioneering work with anti-gas liquids that prevent bloat, gassiness, discomfort and embarrassment.

The Beano information centre is on the Internet at www.beano.net. Beano is on store counters and shelves.

Alan Kligerman saw humanity in distress. With a little research, he found a way to prevent some of the misery.

With a few exceptions, most of them male and young, no one likes the increase in gas production that follows a meal of beans. Alan Kligerman saw a way to attack the problem at its source.

He had previously solved a problem akin to this one, and, in so doing, had become a wealthy man. Kligerman had devised a product called Lactaid which enabled lactose-intolerant people – those who physically cannot digest the particular sugar in milk – to drink milk without suffering a consequent blowout.

Lactaid essentially has a single ingredient – an enzyme called lactase. When Kligerman added lactase to milk, the lactase digested the milk sugar. He began selling it in 1973. Repairing to the library some fifteen years later, he made the discovery that launched the Beano revolution.

He found, buried not so very deep in the scientific literature, an alpha-galactosidase enzyme that helps break down the complex sugars found in many famously gassy foods. The alpha-galactosidase converts them into simpler sugars that the body can handle more comfortably, producing less gas.

Kligerman offered Beano in both solid and liquid form. Three Beano tablets with the first bite of food do the trick for a meal, as do, alternatively, five drops of liquid Beano. The precise formula is 1 Beano tablet or five drops of Beano liquid per serving of problem food (where one serving = $\frac{1}{2}$ cup). A typical meal consists of two to three servings.

For counteracting the forces of bloat, gassiness, dis-

Alan Kligerman, the inventor of Beano, accepts his Ig Nobel Prize from Nobel Laureate Dudley Herschbach. (Roland Sharrillo/*Annals of Improbable Research*)

comfort and embarrassment, Alan Kligerman won the 1991 Ig Nobel Prize in the field of Medicine.

The winner travelled from his company headquarters in New Jersey to the Ig Nobel Prize Ceremony. Alan Kligerman graciously accepted the award and told the audience a little about Beano, and more than a little about his newest product, a version of Beano for dogs, sold under the brand name CurTail. He gave the basic sales pitch for CurTail: 'At last an end to those dreadful moments in your living room when people don't know whether to look at the dog or at each other. Would you like your pet to have a new air of innocence so you'll all breathe easier? Try CurTail Drops.'

He then presented Beano to each of the Nobel Laureates and instructed them on its use.

Kligerman's buoyant enthusiasm raised clouds of excitement throughout the room. He left the stage to a thunderous chant of 'Beano! Beano!'

Alan Kligerman sold the rights to Beano (but not to CurTail) to a larger pharmaceutical company, from which he continues to receive royalties. He stands as a living, relatively gas-free endorsement to the power of science in our daily lives.

IG NOBEL PRIZES 2 — Whence the Gas?

Here, as if you needed it, is Alan Kligerman's list of foods that commonly lead to high gas production.

VEGETABLES
Beets
Broccoli
Brussels sprouts
Cabbage
Carrots
Cauliflower
Corn
Cucumbers
Leeks
Lettuce
Onions
Parsley
Peppers, sweet

GRAINS/CEREALS/SEEDS/NUTS
Barley
Breakfast cereals
Granola
Oat bran
Oat flour
Pistachios
Rice bran
Rye
Sesame flour
Sorghum, grain
Sunflower flour
Wheat bran
Wholewheat flour

LEGUMES
Black-eyed peas
Bog beans
Broad beans
Chickpeas
Field beans
Lentils
Lima beans
Mung beans
Peanuts
Peas
Pinto beans
Red kidney beans
Soya beans

OTHERS
Bagels
Baked beans
Bean salads
Chili
Lentil soup
Pasta
Peanut butter
Soya milk
Split-pea soup
Stir-fried vegetables
Stuffed cabbage
Tofu
Wholegrain breads

Foot Malodour

Short-chain fatty acids from the socks and feet of subjects either with strong foot odour or with weak or no foot odour, were extracted with ethyl ether, and then analysed by gas chromatography/mass spectrometry (GC/MS). Short-chain fatty acids were found in greater amounts from those subjects with strong foot odour... By incubating sweat and lipid from subjects with strong foot odour, we succeeded in reproducing the foot malodour.

— from the report by Kanda, Yagi, Fukuda, Nakajima, Ohta and Nakata

The Official Citation

The Ig Nobel Medicine Prize was awarded to:

F. Kanda, E. Yagi, M. Fukuda, K. Nakajima, T. Ohta and O. Nakata of the Shiseido Research Centre in Yokohama, for their pioneering research study 'Elucidation of Chemical Compounds Responsible for Foot Malodour', especially for their conclusion that people who think they have foot odour do, and those who don't, don't.

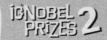

Elucidation of chemical compounds responsible for foot malodour

F. KANDA, E. YAGI, M. FUKUDA, K. NAKAJIMA, T. OHTA AND O. NAKATA

Shiseido Research Center, 1050 Nippa-cho, Kohoku-ku, Yokohama 223, Japan

Accepted for publication 28 December 1989

SUMMARY

Short-chain fatty acids from the socks and feet of subjects either with strong foot odour or with weak or no foot odour were extracted with ethyl ether, and then analysed by gas chromatography/mass spectrometry (GC/MS). Short chain fatty acids were found in greater amounts from those subjects with strong foot odour. Iso-valeric acid was present in all the subjects with

Their study was published in the *British Journal of Dermatology*, volume 122, number 6, June 1990, pp 771–6.

When one's feet smell unpleasant, it is polite to wonder why. F. Kanda, E. Yagi, M. Fukuda, K. Nakajima, T. Ohta and O. Nakata pursued this interest more thoroughly than mere politeness would dictate.

Kanda, Yagi, Fukuda, Nakajima, Ohta and Nakata all work at the Shiseido Research Centre in Yokahama, Japan. None of them is purely a foot-malodour specialist. In their search for chemical compounds responsible for foot malodour, the six hoped to reap the intangible benefits of interdisciplinary collaboration.

Their investigation had three phrases.

In Phase 1, they assembled a panel of sniffers – ten men and ten women, all between 20 and 35 years old, who had not been specially trained to detect foot malodour or any of the other odours that would be included in the experiment.

The scientists mixed eight different potions, each containing chemicals of which they were suspicious, chemicals known to often lurk in other fragrant body parts – armpits and vaginas and scalps. After the sniffers sniffed each potion, the scientists asked them if the smell was familiar, and if it was, to say whether it resembled foot odour, armpit odour, or something else.

The sniffers all agreed that the potions smelled more or less like foot or armpit, but disagreed as to which, and how closely.

In Phase 2 of the investigation, Kanda, Yagi, Fukuda, Nakajima, Ohta and Nakata then recruited five healthy men whom they considered to have strong foot odour, and another five whom they considered to have little or none. They verified these considerations by having all the men exercise vigorously for thirty minutes. Then the scientists took the men's socks and placed those socks inside a laboratory apparatus for five hours, to produce what they term 'sock extract'.

In the third and final phase of the investigation, the scientists tried to reproduce the malodour using various combinations of the chemicals that had earlier been sniffed by the panel of sniffers. They took these artificial foot-odour concoctions and compared them with the genuine sock extracts. The comparison was done with a Hitachi M-80 mass spectrometer via a Hewlett Packard 5710A gas chromatograph.

The result? They identified several different

chemicals, most of them short-chain fatty acids, that seem to be present in or on smelly feet. And they identified one special chemical – iso-valeric acid – that seems to play a suspiciously large role. That is as far as they got. The full story of the chemistry of foot malodour has yet to be found, but Kanda, Yagi, Fukuda, Nakajima, Ohta and Nakata took an historic step in its general direction.

Perhaps more important, in an immediate sense, they were able to confirm, vis-à-vis the ten men from whom they took sock extracts, that those who thought they had foot odour did, and those who didn't, didn't.

The research did not go unnoticed or unappreciated. Part of it, described in a separate paper titled 'Elucidating Body Malodour to Develop a Novel Body Odour Quencher', was recognized with an award in 1988 from the International Federation of Societies of Cosmetic Chemists.

Later, for their fundamental, if preliminary, discoveries about smelly feet, F. Kanda, E. Yagi, M. Fukuda, K. Nakajima, T. Ohta and O. Nakata won the 1992 Ig Nobel Prize in the field of Medicine.

The winners could not, or would not, attend the Ig Nobel Prize Ceremony.

10. Food Adventure and Aftermath

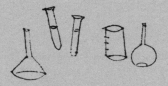

Food holds a fascination for anyone who prepares or uses it. This chapter presents four food-related Ig Nobel Prize-winning achievements, of varying flavours:

The Spiceless Jalapeño Chile Pepper

Spam

🌱

Collector's Choice

🌱

The Constipated Serviceman

🌱

The Spiceless Jalapeño Chile Pepper

jalapeño, n. Also jalapeño pepper. A very hot green chilli pepper, used esp. in Mexican-style cooking... 1964 MRS L.B. JOHNSON *White House Diary* 25 Apr. (1970) 123 Hash is one of Lyndon's favourite foods, especially with jalapeños. ... 1992 *Vanity Fair* (NY) Mar. 248/1 Predictions make you sweat more than a dose of jalapeño pepper.

— from the *Oxford English Dictionary*, second edition, additions 1993

IG NOBEL PRIZES 2 A Note for the Niggling

There is debate, sometimes fiery, often not, about how to spell the name of the pepper. The American spelling is (usually) 'chile'. The British spelling is (usually) 'chilli'. The tasty stew made of beans and beef and peppers and/or whatever the cook insists on using has a name that sounds the same, but is generally spelled, at least in America, 'chili'.

The Official Citation

The Ig Nobel Biology Prize was awarded to:

Dr Paul Bosland, director of the Chile Pepper Institute, New Mexico State University, Las Cruces, New Mexico, for breeding a spiceless jalapeño pepper.

IGNOBEL PRIZES 2   A technical description of the pepper was published as '"NuMex Primavera" Jalapeño', Paul W. Bosland and Eric J. Votava, *HortScience*, volume 33, number 6, 1988, p 1085–6.

Jalapeño chile peppers are prized for being fiery-hot. The world's top expert on chile peppers was praised – and here and there cursed – for breeding a variety of jalapeño that has virtually no spice.

Professor Paul Bosland runs the Chile Pepper Institute, an international research centre based at New Mexico State University, just north of the US-Mexico border. Here is his official ten-second guide to chile pepper hotness:

Chiles are hot because they have a compound, or a set of compounds, called capsaicinoids that's found inside the fruit, along the placenta. And contrary to a lot of beliefs, the walls have no heat, the seeds don't have any heat, they're only in one little area inside – where the orange colouring is – that is the capsaicinoids. So the more orange you see there, the hotter the chile.

Of the many varieties of chile pepper, the jalapeño is famous for being one of the hottest. Through years of very hard effort, Professor Bosland bred a variety of jalapeño that has virtually no spice.

This   excitingly   bland,   seemingly   oxymoronic

creation does have some spice – just not much. For a jalapeño, its spiciness is pitiful. The new pepper is named NuMex Primavera. Its very existence strikes some people as tasteless, but it is not literally so. In fact, NuMex Primavera does have lots of taste; that was half the point of breeding it.

Restaurant chefs who prepare chile-pepper-based dishes have always had a problem. The spiciness of any two individual peppers can vary wildly. If three people at a table order the chile rellenos, one of them might get a fairly spicy dish, the second a very spicy dish, and the third a non-fatal-heart-attack dish.

Here, to give some context, is a recipe for chile rellenos.

INGREDIENTS:
6 large chiles, roasted, peeled and sliced lengthwise, with seeds removed, and stuffed with Jack, Muenster or asadero cheese
3 egg whites
2/3 cup milk
1/3 cup white flour
1/3 cup wholewheat flour
1/2 tsp. salt
vegetable oil

DIRECTIONS:
Put all but the egg whites in a blender and mix. Place in medium-size bowl. Beat egg whites. Fold gently into flour mixture. Coat each stuffed chile pod with the batter; cook in hot oil in a skillet until brown. Serve with mole, or red or green chile sauce.

Good chefs want to be able to control the taste and

spiciness of their creations, so Professor Bosland hoped to provide a chile pepper with the true characteristic taste of the jalapeño, but only a consistently tiny amount of the fire. This he achieved. 'The NuMex Primavera jalapeño chile pepper,' he explained triumphantly, 'dilutes the heat without changing the flavour.'

Professor Bosland's other reason for creating this weirdly moderate pepper was simply to show that he could. He is still labouring to breed a version that has 'absolutely no heat'.

There is irony in his damping the eternal flame, for some people, Professor Bosland says, think he has 'sold his soul to the devil'.

For taming the legendary fire of the jalapeño, Paul Bosland won the 1999 Ig Nobel Prize in the field of Biology.

Professor Bosland travelled from Las Cruces, New Mexico, to the Ig Nobel Prize Ceremony at Harvard. In accepting the Prize, he said:

> There is some method behind the madness. I want all of you to become 'chile heads'. So my plan was to make them mild, then you would begin eating them. Then I'll make them a little hotter. You'll eat more. And then, before you know it, you'll be able to eat the really hot ones. I'd like to say in closing, the only thing you have to remember is: chile peppers are like rock 'n' roll; whether you like them or not, they're here to stay. Thank you very much.

Afterwards he passed out a few samples of the new

and the old peppers. Everyone who dared taste them agreed that the difference is impressive.

Spam

As the Spam cans approach the cooker in a line, an arm swings out and pushes twenty-four of them onto a shelf, and the shelf starts moving upward, like a carnival Ferris wheel. During the next two hours, sixty-six thousand Spam cans will travel up and down eleven chambers in the cooker, and along the way they will be heated to the point of sterilization, washed, and cooled. By now you've probably surmised something that surprises people unfamiliar with canning: Spam is cooked in its can.

— from the book *Spam: a Biography*

The Official Citation

The Ig Nobel Nutrition Prize was awarded to:

The utilizers of Spam, courageous consumers of canned comestibles, for 54 years of undiscriminating digestion.

Spam, the pink, mushy, tinned-meat product, is consumed by millions of human beings, some of whom enjoy the experience. Just as there is something indefinable about Spam, there is something unfathomable about the eating of it.

There is no accurate way to determine how many people have eaten Spam. Hormel Foods, the source of all Spam, produced the first can in 1937, and, according to its estimates, the five-billionth in 1994.

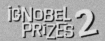

The book *Spam: a Biography*, by Carolyn Wyman, Harcourt Brace & Co., 1999, is a comprehensive source of information about Spam and those who ingest it.

Vegetarians do not consume Spam, at least not knowingly, or, at the very least, not admittedly. Nor do people who forswear pork. Those groups aside, Spam is liable to find its way into the digestive system of almost anyone, anywhere. This simple substance contains pork, ham, salt, water, sugar and sodium nitrate. The sodium nitrate gives the Spam its colour and renders it unpalatable to bacteria, thus giving the concoction the lengthy shelf life which so deliciously characterizes it in the minds of admirers.

The consumers of Spam are a varied lot, yet almost without exception all ingested their first mouthful for one of just three reasons.

For some, it was a matter of economy. Spam is, and always has been, an inexpensive relative of meat.

For others, it was a matter of duress. During World

Many soldiers ingested Spam during World War II. (Photo: US Army)

War II the United States military purchased more than 150 million pounds of Spam. Much was fed to the troops, who were given little alternative, and a considerable amount was sold or given to the peoples of other nations, many of whom were otherwise starving. In fact, much of that 150 million pounds of military Spam was not really Spam – it was a cheap Spam substitute – but most of the people who ate anything of that ilk believed, at least in retrospect, that they were eating Spam. After the war many continued to indulge.

Nearly all other consumers of Spam are children of Spam eaters. Indoctrinated early in life, they became habitually enamoured of the substance.

Collectively, the consumers of Spam consume it

everywhere, in all weathers, and in every conceivable configuration and combination. There is almost no foodstuff that someone, somewhere, has not mixed with Spam and eaten.

Most consumers of Spam acknowledge the basic fact that is its virtue and its vice. No matter how Spam is cooked, no matter what is added to or placed atop it, the dominant taste is always the same.

The consumers of Spam have celebrated it in paint, song, verse, and every other form of artistic expression, but more significantly, they have consumed it, again and again and again.

For passing Spam through their systems so diligently and for so long, the consumers of Spam won the 1992 Ig Nobel Prize in the field of Nutrition.

Not all of the winners could attend the Ig Nobel Prize Ceremony, but many did. Dr John S. Meagher, a US Navy veteran of World War II and a continuous consumer of Spam since 1942, accepted the Prize on behalf of all of them. Dr Meagher's remarks were distinguished by his insistent eating of a Spam sandwich while talking. This rendered most of his words undistinguishable, but the concluding phrase rang out loud and clear: 'When you talk about family values, you're really talking about Spam.'

Collector's Choice

For one reason or another, I have always been interested in useless matters.
— Arvid Vatle

The Official Citation

The Ig Nobel Medicine Prize was awarded to:

Dr Arvid Vatle of Stord, Norway, for carefully collecting, classifying and contemplating which kinds of containers his patients chose when submitting urine samples.

Unyttig om urinprøver

Forfattaren har i løpet av 12 månader registrert typen av emballasje som pasientane har bringa urinprøvene til hans allmennpraksis i. Artikkelen gjer ikkje krav på originalitet, og forfattaren ønskjer heller ikkje at han skal bli vurdert meritterande i noko fag. Men arbeidet har vore til stor glede for forfattaren.

Uroskopi, alkymi og astrologi var tre hovudfundment i moderne diagnostikk gjennom mesteparten av mellomalderen. Urinen var rett nok ikkje ei av dei fire klassiske kroppsvæskene i humoralpatologien, men likevel vart urinprøver brega til lækjaren, som vurderte mellom anna farge og sediment.

I perioden mai 1997 – mai 1998 engasjerte eg meg i følgjande, totalt nyttelause studie: eg registrerte 164 urinprøvebehaltarar med ein variasjon på 110 ulike av dei (tab 1).

Einaste kriterium for seleksjonen var at urinprøva vart levert i ein emballasje som ikkje var av den fordjipkjede, sterile typen.

Det visste seg at det ikkje tydde noko for urinanalysen om prøva var på ei halvliters Cola-flaske eller på eit tomatpurgllas, for emballasjen var alltid svært godt reingjort.

Artikkelen er ikkje meint å vere medisinhistorisk, med finjer frå den epoken sin uroskopi. Nei, han er rett og slett skriven i stor

Bells Old Scotch Whisky og Koskenkorva vodka, må ein halde seg stramt i tinamane: ein slik observasjon treng ikkje tyde at personen har fått eller alltid har hatt hang til flaska. Det treng således ikkje vere eit ordlaust rop om hjelp!

Det treng heller ikkje vere ein demonstrasjon mot helsevesenet sine åtvaringar mot alkohol og andre nytingsmiddel. Det kan vere reine dumpen at den personen treng bo driftsleioren til gonger valde slik emballasje.

Det same må hovdaut for emballasje som har innehalde Q10, purpumolhalt eller vitamin E.

Vlktige signal?
Eller er dette kanskje likevel signal som er verdt å legge merke til? Vil nokon meddele noko når dei bringar

His study was published as 'Unyttig om Urinprøver', Arvid Vatle, *Tidsskift for Den Norske Laegeforening* (the journal of the Norwegian Medical Association), number 8, 20 March 1999, p 1178.

A family physician on a small island in western Norway decided to keep track of some information he had previously ignored. Whenever a patient brought in a urine sample, the doctor took note of what kind of container the patient used. Eventually, this resulted in one of the most original reports in the annals of medicine.

Dr Arvid Vatle lives and practises medicine on the island of Stord, Norway, (latitude: 59.86° N, longitude: 5.41°), where the air is clean, the people are on the whole healthy, and the weather is often conducive to reflection. Dr Vatle is of the old school, indulging both a talent for medicine and a fondness for literature. After it occurred to him that no one had ever collected good information about the urine-container preferences of Norwegian medical patients, it was not a far leap to consider that one day, with sufficient amounts of data in hand, he might submit a formal write-up to a reputable medical journal.

During the next year, whenever a patient brought him a urine sample, Dr Vatle took notes about the container. After an intensive twelve months of collecting, he wrote his report and submitted it for publication in *Tidsskift for Den Norske Laegeforening*, the journal of the Norwegian Medical Association. The article is written with deft panache, in Norwegian.

Most of the container types are listed only once in the survey. The most popular kind of container, a particular brand of tomato-puree jar, appeared seventeen times. The variety and range impress themselves on even the casual reader, and even on the reader who is illiterate in Norwegian:

Apocillin 660 milligrams
Asbach Uralt miniflaske
Bergensk Brystbalsam
Coca-Cola Light 0.5 litre
Eggbit gaffelbiter
Fruitopia
Hapa Nestlé
Hindu Kanel
Mills Peanut Butter,
    Grov type
Nitroglycerin 0.5
    milligrams
Normorix mite
Olden naturlig mineralvann
    m kullsyre 0.5 litre
Skin Renew Crem 24 hours,
    60 millilitres
Tab, Full Cola Taste,
    No Sugar
Taco saus, Cara Fiesta
Toilax 5 milligrams
Tysk dressing

The report's details –
especially the table that
indicates the occurrence of
particular bottle types –
may be useful to medical
students and to planners in
the Norwegian bottling
industry. (S. Drew/*Annals
of Improbable Research*)

And many more – over a hundred types altogether.

For studying a neglected aspect of the healer's art,
Dr Arvid Vatle won the 1999 Ig Nobel Prize in the field
of Medicine.

Arvid Vatle and his wife Gillian travelled from Stord,
Norway to the Ig Nobel Prize Ceremony, their trip partly
subsidized by the Norwegian Medical Association. In
accepting the Prize, Dr Vatle said:

> Ladies and gentlemen, with tears in my eyes, I have
> to admit that my present study is not a serious
> exercise in the promotion of science. Nor, for use in
> my career, in any field, be it urology, psychiatry,
> folklore, marketing, or medical history. But, as a
> medical country practitioner, I sometimes find it
> necessary to whistle that I may not weep. However,
> I have to state that further research is necessary to
> survey possible geographic variations in the national
> and even international use of different types of
> containers for urine samples.

IGNOBEL PRIZES 2 — Urine Containers at Harvard Medical School

The day after receiving his Ig Nobel Prize, Dr Vatle lectured at Harvard Medical School on the subject of his research. Here is a much-condensed version of that talk.

Formerly, I remember, stool samples were brought in anything from match boxes to far more voluminous jars.

What does the average patient do when he or she is asked to bring a urine sample? A frantic search will begin, in the cupboards of the kitchen, the bathroom and the sitting room. All the bottles registered in my study have preciously contained food or spices, medicines, perfumes, or strong liquors. You take what you find, no matter what the size and label is. You cleanse it, boil it if possible, and dry it.

Then there is a new challenge: How do you get the urine sample into the bottle if the opening is like the pin point opening of some perfume or liquor bottles? Well, I really don't know, but I am indeed impressed by the degree of precision that some persons seem to have exercised during the operation.

What a boring contrast the ready-made, sterile, plastic containers of hospitals, institutions! There you need no imagination; the translucent wrapping-up container leaves no doubt as to the content.

I kept the study going for twelve months, from May 1997, to May 1998. Finally, on my list there was a variety of 110 different types of containers, totalling 164 samples. All containers produced specially for urine samples were excluded from my study.

What, if anything, does the type of container tell the doctor about the patient?

A respected, elderly gentleman of high standing in the community brought his first urine sample in a bottle labelled Bell's Old Scotch Whisky. His second appeared in a bottle of Koskenkorva Finnish vodka.

Was this honourable gentleman addicted to the bottle? Or did he indirectly invite me for a drink on an occasion? Did he hold shares in the companies producing the liquors? Or did he want to demonstrate defiance against the paternalistic anti-alcohol warnings of the medical community and the ministry of health in Norway? I don't know the answer. The bottles probably just happened to be available in his home.

Does a urine sample brought in a bottle having contained antidepressant tables communicate that the patient is in need for help and comfort?

One of the most astonishing containers was a Mum roll-on deodorant. The ball had been taken away and the deodorant replaced by a malodorant.

Some samples are brought in delicate plastic or cardboard containers, usually well wrapped up in large plastic bags from the local supermarket.

By far the most frequently used container in my general practice was the jar of Stavland's tomato puree. Stavland is a producer on the neighbouring island of Bømlo.

Why, I wonder, is Stavland's tomato-puree jar the most frequently used container for urine samples in my practice? After deep thought, I concluded that the explanation probably would be found in the widespread use of Stavland's tomato puree in the households of our region. Such is also the case with Stavland's Genuine Heather Honey, which was the third most frequently used container. In addition, they have a wide opening and a lid that shuts well.

The Constipated Serviceman

CONCLUSION: The prevalence of constipation in US servicemen deployed in a field environment has been determined to be 30.2% based on infrequent bowel movements and 34.1% based on presence of symptoms of constipation. This significantly exceeds the prevalence of constipation while these men are at home... Consideration should be given to evaluating measures to improve bowel function in the field.

— from the report by Sweeney, Krafte-Jacobs, Britton and Hansen

The Official Citation

The Ig Nobel Biology Prize was awarded to:

W. Brian Sweeney, Brian Krafte-Jacobs, Jeffrey W. Britton

and Wayne Hansen, for their breakthrough study, 'The Constipated Serviceman: Prevalence Among Deployed US Troops,' and especially for their numerical analysis of bowel-movement frequency.

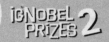

The Constipated Serviceman: Prevalence among Deployed U.S. Troops

LCDR W. Brian Sweeney, MC USNR *　　　LCDR Jeffrey W. Britton, MC USNR‡
Brian Krafte-Jacobs, MD†　　　　　　　LT Wayne Hansen, NC USN§

The prevalence of constipation in deployed servicemen was determined in a sample of military personnel aboard the USS Iwo Jima LPH 2 during Operation Desert Shield. Results were obtained from a bowel function questionnaire issued to 500 deployed marines and sailors. When constipation is defined as no bowel movement for greater than 3 days, 3.9% of the Marine/sailor personnel are constipated when in their home [...] obtained, and each of three pages asked the same meal- and bowel function-related questions for three environments—home, ship, and field. Specifically asked was the number of meals eaten daily, the frequency of bowel movement, and the consistency of bowel movements (choices being liquid, soft, firm, and rock hard). In addition, participants responded (by circling almost never, sometimes, or always) to questions con[...]

Their study was published in *Military Medicine*, volume 158, August, 1993, pp 346–8.

The full nature and extent of constipation among combat troops was not measured with high precision until 1993, when W. Brian Sweeney, Brian Krafte-Jacobs, Jeffrey W. Britton and Wayne Hansen completed their study of what is a very hard problem.

Military medical personnel frequently see occurrences of constipation or diarrhoea.

Diarrhoea during deployment periods has been frequently studied and described. See, for example, Rudland, Little, Kemp, Miller and Hodge's report 'The Enemy Within: Diarrheal Rates Among British and Australian Troops in Iraq' (published in *Military Medicine*, volume 161, number 12, December 1996, pp 728–31), wherein it is reported that: 'The aim was to document different rates of diarrhoea in British and Australian troops. The British ... experienced higher rates of diarrhoea (69% of British troops compared with 36% of Australian troops).'

However, constipation rates among warriors has been neither so well studied nor so well understood as diarrhoea rates. A particularly vexing question nagged at both the troops and their care givers.

Is the incidence of constipation different when troops are actually in the field than it was while they were travelling to the general theatre of combat? W. Brian Sweeney, Brian Krafte-Jacobs, Jeffrey W. Britton and Wayne Hansen set out to answer this question.

The researchers drew up a bowel-function questionnaire, which was then issued to American marines and sailors engaged in Operation Desert Shield in Iraq, in 1991.

The three-page questionnaire asked a variety of bowel-function-related questions for each of three environments – at home, on the ship and in the field. The subjects were asked about the number of meals eaten daily, the frequency of bowel movements, and the consistency of the bowel movements, the choices being liquid, soft, firm and rock hard. They were also asked questions concerning straining, pain and other difficulties, the choices here being never, sometimes, or always.

The questionnaires were distributed as the marines and sailors waited in the evening-meal line. Five hundred individuals agreed to answer the survey, provided they could do it anonymously. Three individuals refused to participate under any conditions.

The researchers tabulated the results. Here is the raw data that emerged:

BOWEL-MOVEMENT FREQUENCY WHILE AT HOME, WHILE ON THE SHIP AND WHILE IN THE FIELD

Days Between Bowel Movements	At Home	On Ship	In the Field
<1 day	4.5%	2.0%	1.0%
1–2 days	70.6%	55.5%	22.6%
2–3 days	21.0%	36.4%	46.2%
4–5 days	3.1%	5.2%	23.6%
6–10 days	0.8%	0.8%	6.1%
>10 days	0%	0%	0.5% (2 individuals)

(Note: In their formal report, the researchers present some of this data in units of 'bowel movements per day'.)

When Sweeney, Krafte-Jacobs, Britton and Hansen analysed their data, they found much to chew upon. Here are some of the more important observations:

> When constipation is defined as no bowel movement for greater than 3 days, 3.9% of the marine/sailor personnel are constipated when in their home environment as compared to 6.0% when they are aboard ship and 30.2% while in the field.
>
> Alternatively, when constipation is defined as the presence of certain anorectal complaints (hard stools, straining, painful defecation...), the incidence is 7.2% when at home as compared to 10.4% aboard ship and 34.1% in the field.

This was startling. When they were on the ship, the men's bowel-movement frequency was essentially the same as when they were at home. But when they left the ship to go to the battlefield, things tightened up.

The researchers recognized the severity of the situation. They suggested that efforts be made to ease the situation. They minced no words in recommending that further research be done:

> Significantly more servicemen will be constipated when in the field as compared to their home environment. Since approximately one-third of navy/marine corps personnel deployed in a field environment will be constipated, preventive measures ought to be evaluated.

For their efforts to bring relief to the distressed fighting man and woman, W. Brian Sweeney, Brian Krafte-Jacobs, Jeffrey W. Britton and Wayne Hansen won the 1994 Ig Nobel Prize in the field of Biology.

Dr Brian Sweeney travelled to the Ig Nobel Prize Ceremony. Dr Sweeney had by that time retired from the Navy and taken a job at the University of Massachusetts Medical Center in Worcester, Massachusetts, so his journey was not a lengthy or costly one. In accepting the Prize for himself and his colleagues, he said:

> I would like to acknowledge all of our wonderful US servicemen who are willing to become constipated for their country ... I initially felt that [constipation] was related to diet – low fibre, or eating those MREs [Meals, Ready to Eat], or poor water intake. Until one of the marines in the field said to me, 'Doc, let me tell ya. When we're out in the field, we're scared shitless.'

IG NOBEL PRIZES 2 What is an MRE?

The constipated American military personnel who were the subjects of Sweeney et al.'s report met much of their nutritional need by consuming MREs. MRE stands for 'Meals, Ready to Eat'. Here are some official facts about MREs. This information is from the US Defence Logistics Agency.

WHAT IS IT?

The Meal, Ready-To-Eat (MRE) is designed to sustain an individual engaged in heavy activity such as military training or during actual military operations when normal food service facilities are not available. The MRE is a totally self-contained operational ration consisting of a full meal packed in a flexible meal bag.

WHAT IS IN IT?

There are twenty-four different varieties of meals (one of which, Menu #10, is described below). Components are selected to complement each entrée as well as provide necessary nutrition.

HOW LONG WILL IT LAST?

The shelf life of the MRE is three years at 80 degrees F.

IgNOBEL PRIZES 2 HOW CAN I ORDER IT?
The National Stock Number for a case of MREs (12 MREs per case) is 8970-00-149-1094.

MENU #10
Chili and Macaroni
Pound Cake
Peanut Butter
Wheat Snack Bread
Cocoa
Beverage Powder
Hot Sauce
Accessory Packet, E
Spoon
Flameless Heater (Note: Not included in MREs manufactured prior to 1992)

ACCESSORY PACKET COMPONENTS
(Packet E)

Tea	Chewing Gum
Sugar	Matches
Creamer	Toilet Tissue
Salt	Hand Cleaner

11. It's Brains You Want

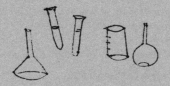

Brainy folks who are curious about brains point out that brains are most curious. Here are three curious examples of brain-related projects that won Ig Nobel Prizes:

The Brains of London Taxi Drivers

The Effects of Chewing-gum Flavour on Brain Waves

The Intelligence of Single-nostril Breathing

The Brains of London Taxi Drivers

Structural MRIs (Magnetic Resonance Images) of the brains of humans with extensive navigation experience, licensed London taxi drivers, were analysed and compared with those of control subjects who did not drive taxis.

— from the published report by Eleanor Maguire and colleagues

The Official Citation

The Ig Nobel Medicine Prize was awarded to:

Eleanor Maguire, David Gadian, Ingrid Johnsrude, Catriona Good, John Ashburner, Richard Frackowiak and Christopher Frith of University College London, for presenting evidence that the brains of London taxi drivers are more highly developed than those of their fellow citizens.

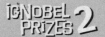

Navigation-related structural change in the hippocampi of taxi drivers

Eleanor A. Maguire*†, David G. Gadian‡, Ingrid S. Johnsrude†, Catriona D. Good†, John Ashburner†, Richard S. J. Frackowiak†, and Christopher D. Frith†

†Wellcome Department of Cognitive Neurology, Institute of Neurology, University College London, Queen Square, London WC1N 3BG, United Kingdom; and ‡Radiology and Physics Unit, Institute of Child Health, University College London, London WC1N 1EH, United Kingdom

Communicated by Brenda Milner, McGill University, Montreal, Canada, January 28, 2000 (received for review November 10, 1999)

Structural MRIs of the brains of humans with extensive navigation experience, licensed London taxi drivers, were analyzed and compared with those of control subjects who did not drive taxis. The posterior hippocampi of taxi drivers were significantly larger relative to those of control subjects. A more anterior hippocampal region was larger in control subjects than in taxi drivers. Hippocam- *a priori* regions of interest. The data were also analyzed by using a second and completely independent pixel-counting technique within the hippocampus proper. Comparisons were made be- tween the brain scans of taxi drivers, who had all acquired a significant amount of large-scale spatial information (as evi- denced by passing the licensing examinations), and those of a

Their study was published as 'Navigation-related Structural Change in the Hippocampi of Taxi Drivers', *Proceedings of the National Academy of Sciences*, volume 97, number 8, 11 April 2000, pp 4398–403. Also see their several subsequent publications on related topics.

Visitors to London are sometimes shocked by the city's taxi drivers. It's impressive that anyone can reliably find his way through the vast, muddled tangle of streets – and the often-not-corresponding muddled tangle of street names.

Impressive, too, is the variety and volume of opinions that flow from the lips of the taxi drivers. There is widespread belief that, ton for ton, the London cabby's personality is larger than anyone else's. As the twenty-first century was arriving, scientists and doctors at University College, London, produced evidence that, inside a cabby's skull, it's not just the personality that's oversized.

Why would anyone study the brains of taxi drivers, rather than say, rocket designers, stamp collectors, or aficionados of horse racing? The Maguire/Gadian/Johnsrude/Good/Ashburner/Frackowiak/Frith report explains:

> Taxi drivers in London must undergo extensive training, learning how to navigate between thousands of places in the city. This training is colloquially known as 'being on the Knowledge' and takes about two years to acquire on average. To be licensed to operate, it is necessary to pass a very stringent set of police examinations. London taxi drivers are therefore ideally suited for the study of spatial navigation. The use of a group of taxi drivers with a wide range of navigating experience permitted an examination of the direct effect of spatial experience on brain structure.

So there was good reason to suspect that, inside a cabby's head, there is something unusual. How to find that certain something, though, was not obvious.

Maguire and her team went about it methodically.

They had reason to believe that one particular part of the brain – the colourfully named hippocampus – is heavily involved when a person figures out how to navigate from one place to another.

They had at their disposal a modern tool, the not-so-colourfully named Magnetic Resonance Imaging (MRI) scanner. Used properly, the scanner can show fine-grained detail of the innards of a living person's brain.

Maguire, Gadian, Johnsrude, Good, Ashburner, Frackowiak and Frith got hold of sixteen taxi drivers – all of whom had acquired The Knowledge – and took MRI scans of their brains. They then compared, painstakingly, in great detail, the taxi-driver brain scans with brain images of sixteen non-taxi drivers.

What did they find? The brains looked just about identical – *except* in the hippocampus.

Even non-scientists may wish to savour a few of the hippocampus-related snippets from their technical report:

> [We saw] evidence of regionally specific structural differences between the hippocampi of licensed London taxi drivers compared with those of control subjects. Taxi drivers had a significantly greater volume in the posterior hippocampus.
>
> [Our results] indicate that the professional dependence on navigational skills in licensed London taxi drivers is associated with a relative redistribution of gray matter in the hippocampus.
>
> Right hippocampal volume correlated with the amount of time spent as a taxi driver...

Maguire, Gadian, Johnsrude, Good, Ashburner, Frackowiak and Frith go on to say that this accords well with previous studies done on rodents and monkeys. Then they sum it all up:

> A basic spatial representation of London is established in the taxi drivers by the time the Knowledge is complete. This representation of the city is much more extensive in taxi drivers than in [other people]. Among the taxi drivers, there is, over time and with experience, a further fine-tuning of the spatial representation of London, permitting increasing understanding of how routes and places relate to each other. Our results suggest that the 'mental map' of the city is stored in the posterior hippocampus and is accommodated by an increase in tissue volume.

What they found, really, was that compared to other people, each of these taxi drivers had a little more of the brain's famous 'gray matter' in the front of the hippocampus, and a little less of it in the back of the hippocampus.

After the study was published in a scientific journal, it became sensational news. The press eagerly reduced it to headlines:

Being a cabbie broadens the mind
Daily Telegraph

Taxicology
Economist

Cabbies grow bigger brains to store all the Knowledge
Daily Mail

WHY YOUR CABBY'S GOT A BIG 'UN
That's his brain, folks!
Daily Sport

London's cabbies, those who deigned to comment, expressed a diversity of opinion:

> The job's definitely mentally challenging ... although I've never felt my brain expand.
>
> My wife certainly wouldn't agree my brain has grown.
>
> We've got big mouths, but not big brains.

For exploring the labyrinthine brains of London taxi drivers, Eleanor Maguire, David Gadian, Ingrid Johnsrude, Catriona Good, John Ashburner, Richard

Eleanor Maguire, the lead scientist on the brains-of-London-taxi-drivers study, prepares to receive her Prize from Nobel Laureate Dudley Herschbach (wearing white coat). Ig Nobel Minordomo Juliet Lunetta uses a small electric fan to help Dr Maguire keep her cool. (Margaret Hart/*Annals of Improbable Research*)

Frackowiak and Christopher Frith were awarded the 2003 Ig Nobel Prize in the field of Medicine.

Eleanor Maguire travelled to the Ig Nobel Ceremony. In accepting the Prize, she told the audience:

> People often say to me, 'Never mind the science. Your work really just means you get a lot of free rides in taxis.' Unfortunately that's never happened – until today. For the first time, in Cambridge [Massachusetts], I got into a taxi, was chatting with the taxi driver, and he gave me fifty per cent off my taxi fare for showing that taxi drivers

David Gadian shows the good citizens of Manchester some MRI images of the brains of London taxi drivers. This photo was taken by Ig Nobel Biology Prize winner C.W. Moeliker (of 'Homosexual Necrophilia in the Mallard Duck' fame) during the 2004 Ig Nobel Tour of the UK and Ireland.

are special. So I'm grateful to Cambridge for that and for this wonderful prize. Thank you.

In later talks to an eager public, during the Ig Nobel Tour of the UK and Ireland, team members pointed out that intriguing, large questions still loom. Among them: does learning about London really twist the innards of a person's hippocampus? Or could it be that he who becomes a London taxi driver already has a peculiar hippocampus?

As of this writing, much remains to be studied, there in the depths and shallows of the brains of London taxi drivers.

The Effects of Chewing-gum Flavour on Brain Waves

Global complexity of spontaneous brain electric activity was studied before and after chewing gum without flavor and with 2 different flavors... Brain electric activity was assessed through Global Omega (Ω)-Complexity and Global Dimensional Complexity (GDC)... Global Omega-Complexity appears to be a sensitive measure for subtle, central effects of chewing gum with and without flavor.

— from the published report by Yagyu, Wackermann, Kinoshita, Hirota, Kochi, Kondakor, Koenig and Lehmann

The Official Citation

The Ig Nobel Biology Prize was awarded to:

T. Yagyu and his colleagues from the University Hospital of Zurich, Switzerland, from Kansai Medical University in Osaka, Japan, and from Neuroscience Technology Research in Prague, Czech Republic, for measuring people's brain-wave patterns while they chewed different flavours of gum.

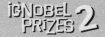

Chewing-Gum Flavor Affects Measures of Global Complexity of Multichannel EEG

··
Abstract
Global complexity of spontaneous brain electric activity was studied before and after chewing gum without flavor and with 2 different flavors. One-minute, 19-channel, eyes-closed electroencephalograms (EEG) were recorded from 20 healthy males before and after using 3 types of chewing gum: regular gum containing sugar and aromatic additives, gum containing 200 mg theanine (a constituent of Japanese green tea), and gum base (no sugar, no aromatic additives); each was chewed for 5 min in randomized sequence. Brain electric activity was assessed through Global Omega (Ω)-Complexity and Global Dimensional Complexity (GDC), quantitative measures of complexity of the

Their study was published as 'Chewing-Gum Flavor Affects Measures of Global Complexity of Multichannel EEG', T. Yagyu et al., *Neuropsychobiology*, volume 35, 1997, pp 46–50.

An international team of brain scientists recorded and analysed people's brain waves while those people chewed three different flavours of chewing gum.

Dr Takami Yagyu and his scientific collaborators, Drs Wackermann, Kinoshita, Hirota, Kochi, Kondakor, Koenig and Lehmann published a detailed, technically complex report explaining exactly what they had done, and why.

The researchers recruited twenty men to chew the gum. Each volunteer had to chew three different wads of gum. The first contained no sugar or other flavourings. The second was Relax Gum, a brand popular in Japan. The final wad was Relax Gum with an added ingredient – a heavy dose of green-tea flavouring, also popular in Japan.

The experimenters took care to see all the volunteers used the same gum-chewing technique, and that their chewing was carried out under consistent, optimal conditions.

Each chewer had eleven electrodes affixed to his head while he chewed, and had to follow a strict five-step protocol.

First, he had to open his eyes for twenty seconds. Then he had to close his eyes.

Then, for a period of exactly 60 seconds, he had to avoid chewing gum. During this time, the experimenters recorded electrical activity from his brain on an elec-troencephalograph.

Next, he had to chew without stopping for 300 seconds, while the experimenters recorded further electrical activity from his brain.

After that, he had to avoid chewing, again for 60 seconds. The experimenters recorded this on the encephalograph too.

At this point, he got a ten-minute pause, without gum.

He then had to repeat the whole process twice, each time with a different flavour of gum.

What was the point? Why did Dr Yagyu and his colleagues do this experiment? They had two motiv-ations. First, they say, they had noticed that technology now makes it possible to measure many things in new, complicated ways. Or as they described it: 'Developments in theoretical physics during the last decades, particularly the theory of non-linear dynamical systems (chaos theory), provided new methods for the assessment of the complexity of bio-electric signals.'

And, second, they had made the scientific observation that – in their words – 'Gum chewing is a very popular pastime of the general population.'

In short: they did it to demonstrate it could be done.

What did they discover? The obvious, of course. As they put it: '[this study] has shown that Global-Omega-Complexity can detect subtle changes in the functional state of the brain as expressed in its electrical activity.' In more technical language, that means: yes, they *could* measure people's brain waves while those people chewed three different flavours of chewing gum.

For their scientific interest in gum chewing and brains, Dr Takami Yagyu and his colleagues won the 1997 Ig Nobel Prize in the field of Biology.

The winners could not, or would not, attend the Ig Nobel Prize Ceremony.

The following year, the group carried their work a step further, investigating not only the effects of the gum's flavour, but also the effects of its smell. This is an example of how science progresses, a step at a time.

The Intelligence of Single-nostril Breathing

This paper correlates uninostril airflow with varying ratios of verbal/spatial performance in 23 right-handed males. Relatively greater cognitive ability in one hemisphere corresponds to unilateral forced nostril breathing in the contralateral nostril. Cognitive performance ratios can be influenced by forcibly altering the breathing pattern.

— from the published report by Buebel, Shannahoff-Khalsa and Boyle

The Official Citation

The Ig Nobel Medicine Prize was awarded to:

Marcia E. Buebel, David S. Shannahoff-Khalsa and Michael R. Boyle, for their invigorating study entitled 'The Effects of Unilateral Forced Nostril Breathing on Cognition'.

IGNOBEL PRIZES 2

THE EFFECTS OF UNILATERAL FORCED NOSTRIL BREATHING ON COGNITION

DAVID S SHANNAHOFF-KHALSA

Senior Staff Laboratory, The Salk Institute for Biological Studies, P.O. Box 85800, San Diego, California 92138

MICHAEL R. BOYLE

Peptide Biology Laboratory, The Salk Institute for Biological Studies, P.O. Box 85800, San Diego, California 92138

and

MARCIA E. BUEBEL

Department of Psychology, Catholic University, Washington, D.C.

(Received October 18, 1990)

Ultradian rhythms of alternating cerebral dominance have been demonstrated in humans and other mammals during waking and sleep. Human studies have used the methods of psychological testing and electroencephalography (EEG) as measurements to identify the phase of this natural endogenous rhythm. The periodicity of this rhythm approximates 1.5 - 3 hours in awake humans. This cerebral rhythm is tightly coupled to another ultradian rhythm known as the nasal cycle, which is regulated by the autonomic nervous system, and is exhibited by greater airflow in one nostril, later switching to the other side. This paper correlates uninostril airflow with varying ratios of verbal/spatial performance in 23 right-handed males.

Their study was published in the *International Journal of Neuroscience*, volume 57, 1991, pp 239–49.

When people breathe through just one nostril at a

time – rather than through both simultaneously – they can manipulate how well their brains work. That is the theory that Marcia Buebel, David Shannahoff-Khalsa and Michael Boyle set out to prove.

Buebel, at Catholic University in Washington, and Shannahoff-Khalsa and Boyle at the Salk Institute for Biological Studies near San Diego, based their remarkable theory on an even more remarkably complex and clever theory.

Here, in layman's language, is the thinking behind the experiment. It has three steps.

First, doctors know that, in broad general ways, the left and right sides of the brain each control the opposite side of the body.

Second, some people believe there is something called a 'nasal cycle', in which the right and left nostrils alternate in doing the heavy breathing.

Third, some people believe there is something called a 'mental cycle', in which the right and left sides of the brain alternate in doing the heavy thinking.

Buebel, Shannahoff-Khalsa and Boyle combined these three ideas into a vivid hypothesis – that by breathing through one nostril, a person can control the nasal cycle, which controls the mental cycle. In sum: the nostril is the key to the intellect.

Such is the theory. Testing it required preparation and much hard work.

Buebel, Shannahoff-Khalsa and Boyle conducted experiments on twenty-three men. Each man had special training in how to breathe through one nostril at a time.

Each man performed some tests, during which the researchers measured the air flow through his left and right nostrils.

In the first test, he read short passages of text, and then wrote down as much of it as he could recall. In another test, he looked at groupings of coloured wooden blocks, and then tried to reconstruct the block arrangements from memory.

The men took these tests with tissue paper plugged tightly into their left nostril, then with tissue paper plugged tightly into their right nostril, then without any tissue paper at all. They used a special yoga technique called 'forced single-nostril breathing' (also known as 'forced uninostril breathing').

According to Buebel, Shannahoff-Khalsa and Boyle's theory, breathing though the left nostril stimulates the right side of the brain. And on the other hand (or, as some may say, 'on the other nostril') breathing through just the right nostril stimulates the brain's left side.

It's a stunningly simple theory, and the experiment produced simply stunning results, which the team reported in simple language:

> Our results indicate that uninostril predominance is associated with varying ratios of cognitive performance and also that altering the phases of the nasal cycle by forced breathing in the nondominant nostril can influence cognitive performance.

Thus, Buebel, Shannahoff-Khalsa and Boyle proved to their satisfaction that breathing skilfully through one nostril makes you think, and makes other people wonder.

For elucidating the relationship between nostrils and brains, Marcia E. Buebel, David S. Shannahoff-Khalsa and Michael R. Boyle won the 1995 Ig Nobel Prize in the field of Medicine.

The winners did not attend the Ig Nobel Prize Ceremony. One co-author was eager to come, but was persuaded not to by the other two.

IG NOBEL PRIZES 2 — Idiosyncratic Admiration

A few weeks after the Prize was awarded, the Ig Nobel Board of Governors received the following letter from an admirer:

Dear Idiot:

Thank you so much for slandering an outstanding researcher (Beubel and Shannahoff-Khalsa). So what if a cardiology group at UCLA replicated their research. So what if our group used their methodology and found it applicable to over 17 different physiological parameters. So what if our cardiology group found that their parameter was highly prognostic of coronary artery disease (as done by nuclear medicine testing and power spectral analysis of heart rate variability). After all, what's a few million lives?

So what if not taking the factor they found into consideration INVALIDATES hundreds of prior papers in the cognition literature as well.

The moron who picked their paper for the IGnoble (sic) Prize deserves to have the kind of stuff that Beubel's paper may have prevented (myocardial infarction, glaucoma, hypertension).

You turn my stomach.

Dr Josh Backon

Hebrew University, Jerusalem

Dr Backon, it turns out, is himself the author of a number of research reports about forced single-nostril breathing. These include:

🏆 'Effect of Forced Unilateral Nostril Breathing on Blink Rates: Relevance to Hemispheric Lateralization of Dopamine' (1989).

🏆 'An Animal Analogue of Forced Unilateral Nostril Breathing: Relevance for Physiology and Pharmacology' (1989), in which Dr Backon speculates that 'an animal analogue of this technique could be provided by gluing large plastic pellets in either the right or left nostrils in a group of rats'.
and

🏆 'Forced Unilateral Nostril Breathing: A Technique That Affects Brain Hemisphericity and Autonomic Activity', (1990), in which he reports something rather startling: 'Asymmetrical buttock pressure affects ipsilateral nasal resistance, autonomic tone, and hemisphericity.'

As the closing remark in his letter implies, Dr Backon is also an expert on nausea and vomiting. He published a research paper on that subject in 1991.

12. Much Higher Learning

Kn owledge is said to be a good thing. Some people are said to be in possession of more of this good thing than other people are. This chapter describes six approaches to obtaining, disseminating, or dismissing higher knowledge:

Where the Hell

🙊

High-level Kidnapping

🙊

Evolution Nyet

🙊

The Irrelevance of Understanding

🙊

Dianetics

🙊

The Bible Code

🙊

Where the Hell

Of course, we believers don't need science to validate the truth and 100 per cent accuracy of the Bible. Yet even this very scientific breakthrough and expanded knowledge is predicted in scripture, in the book of Daniel 12:4.

— Jack Van Impe, in the *Van Impe Intelligence Briefing* newsletter

The Official Citation

The Ig Nobel Astrophysics Prize was awarded to:

Dr Jack and Rexella Van Impe of Jack Van Impe Ministries, Rochester Hills, Michigan, for their discovery that black holes fulfil all the technical requirements to be the location of Hell.

Their discovery was announced on the 31 March 2001 television and Internet (www.jvim.com) broadcast of the *Jack Van Impe Presents* programme. The black-hole announcement came at about the 12-minute point of the half-hour-long programme. Videotapes of the broadcast are available from Jack Van Impe Ministries, PO Box 7004, Troy MI, 48007-7004.

By their nature and location, black holes are extremely difficult to study – or so astronomers believed. They first theorized the existence of black holes, then laboriously built their analyses, combining Einstein's theory of general relativity with observations from ground-based optical telescopes, the Hubble Space Telescope, and other precision instruments. But on 31 March 2001, Reverend Jack Van Impe made a simple announcement that changed the way scientists must think about black holes.

Without going into details (television time is expensive and does not always allow for the giving of details), televangelist Jack Van Impe explained that he had examined all the available scientific evidence about black holes. This led him to a simple conclusion: black holes fulfil all the technical requirements to be the location of Hell.

Unlike others in his profession, Reverend Jack Van Impe and his wife Rexella are fervent devotees of science. On their weekly television programme, *Jack Van Impe Presents*, Jack and Rexella (they encourage us to call them by their first names) discuss the latest scientific discoveries. They cull their facts from newspapers and magazines, and interpret them with gusto, vigour, and numbered biblical citations.

Occasionally, Jack and Rexella run across small discoveries that others have overlooked, such as when Jack reported that:

> Christ is first coming to snatch us away in the twinkling of an eye, 1 Corinthians 15:52. Scientists from the General Electric company discovered that 'the twinkling of an eye' is $\frac{11}{100}$ths of a second long.

Rexella Van Impe

Often, they see the importance of facts that, to others, seemed insignificant. Several years before Jack calculated the location of Hell, Rexella reported the location of heaven. Her pinpoint identification was as poetic as it was scientific:

> One of the most inspiring and thrilling of recent disclosures of astronomers is that there is a great empty space in the north in the nebula of the constellation of Orion, a heavenly cavern so gigantic that the mind of man cannot comprehend it and so brilliantly beautiful that words cannot adequately describe it. All astronomers agree there is a huge opening in Orion which is perhaps more than 16,740,000,000,000 miles in diameter. The diameter of the earth's orbit is 186,000,000 miles, which in itself is incomprehensible to man. Yet the opening into this heavenly cavern of Orion is 90,000 times as wide. God's heaven is in the side of the North.

Many of Jack and Rexella's discoveries are incomprehensible to the mind of man.

Jack and Rexella often mention the work of professional scientists. Their chief authority on questions of astronomy is Dr Edgar Lucien Larkin, former director of California's Mt Low Observatory. Dr Larkin's standing in the science community rests on his book *Matchless Altar of the Soul: Symbolized as a Shining Cube of Diamond, One Cubit in Dimensions & Set Within the Holy of Holies in All Grand Esoteric Temples of Antiquity*. The book was published in 1917.

Dr Larkin is also noted for his series of writings on Aryan metaphysics and philosophy. Dr Larkin went to either heaven or hell in 1925.

For their abysmal, far-sighted vision, Jack and Rexella Van Impe won the 2001 Ig Nobel Prize in the field of Astrophysics.

The winners could not, or would not, attend the Ig Nobel Prize Ceremony. The Van Impe's secretary told the Ig Nobel Board of Governors that they had to attend a 'previously scheduled fundraiser'.

Concerned that the Van Impes' contribution to modern astronomy might be under-appreciated by the public, the Board of Governors arranged for MIT astrophysicist Walter Lewin to accept temporary custody of the Prize on behalf of the winners. Here is Professor Lewin's tribute to the Van Impes:

> I accept this prestigious prize on behalf of Jack and Rexella Van Impe for their breakthrough contributions to astrophysics by making the intriguing connection between black holes and hell.
>
> Now, I do real research in black holes. A black hole is one of the most bizarre, exciting, fascinating, enigmatic and mind-boggling objects in our entire universe. Black holes are unimaginable. They go

beyond our wildest expectations, fantasies and dreams. As a scientist, you can't wish for more. Black holes are heaven for us.

However, we were recently straightened out by Jack and Rexella, who have shown that black holes fulfil all the requirements of hell. As a result of their incredible insights, we now have to rethink our ideas about black holes – and I, for one, will be much more careful than ever about getting too close to the inner workings of black holes; and maybe my soul will be spared after all.

I can put this in seven words: Black holes are wonderful, but stay away.

High-level Kidnapping

ALIENS FROM OUTER SPACE ARE NOT HERE TO HELP US!

— headline on the back cover of the book *Secret Life*

The Official Citation

The Ig Nobel Psychology Prize was awarded to:

John Mack of Harvard Medical School and David Jacobs of Temple University, mental visionaries, for their leaping conclusion that people who believe they were kidnapped by aliens from outer space, probably were – and especially for their conclusion that 'the focus of the abduction is the production of children'.

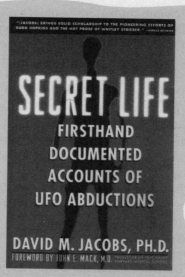

IGNOBEL PRIZES 2

Their study was published in book form as *Secret Life: Firsthand Documented Accounts of UFO Abductions*, Simon & Schuster, 1992.

"[JACOBS] BRINGS SOLID SCHOLARSHIP TO THE PIONEERING EFFORTS OF BUDD HOPKINS AND THE HOT PROSE OF WHITLEY STRIEBER." —KIRKUS REVIEWS

SECRET LIFE

FIRSTHAND DOCUMENTED ACCOUNTS OF UFO ABDUCTIONS

DAVID M. JACOBS, PH.D.

FOREWORD BY JOHN E. MACK, M.D. PROFESSOR OF PSYCHIATRY HARVARD MEDICAL SCHOOL

For years, people joked that creatures from outer space would come to earth and kidnap our women. Dr John Mack and Professor David Jacobs knew it was no joking matter. Braving the inevitable ridicule, they decided to warn the public.

John Mack is a professor of psychiatry at Harvard Medical School. David Jacobs is a professor of history at Temple University. Both approached this subject with deep scepticism.

'When I first became involved with abduction research,' wrote Jacobs, 'it was easy to keep it at arm's length and treat it as an intellectual puzzle. But the more I learned about the abduction phenomenon, the more frightening it became, both personally and in the larger context of its potential effects on society.'

Dr Mack was severely disturbed by what he found. 'For me and other investigators, abduction research has had a shattering impact on our views of the nature of the cosmos.'

With the experienced insight of any good psychiatrist, Dr Mack pointed out that even with this terrible ongoing occurrence there came some good. 'My own work with abductees has impressed me with the powerful dimension of personal growth that accompanies the traumatic experiences.'

Between them, Dr Mack and Professor Jacobs interviewed some three hundred individuals who had tales to tell. Most of them were women, a fact that was not surprising once Dr Mack and Professor Jacobs realized the purpose of the kidnappings.

Dr Mack and Professor Jacobs wrote that the victims tell a consistent and compelling story.

Each victim is taken to a main examining room, which is small and circular. 'Although disoriented, she can still observe what is happening to her.' The aliens remove her clothes and lie her on a table. 'It is eerily quiet.' More often than not, the aliens place restraints on her arms and legs. They examine her body and test her reflexes. They separate her legs and make a quick gynaecological examination. (In the case of a male abductee, they palpate, lift and observe his genitals.) The aliens examine the woman's breasts and insert something into her mouth. Some victims say the examination goes on for years. The aliens implant something small, round and metallic in the abductee's ear, nose, or sinus cavity; or, if this is a return visit, they remove the object. The aliens then do a mindscan, and sometimes go through a bonding ritual which can be sexual in nature. Then they insert a needle and harvest eggs from her belly, or, if this is a return visit, implant an embryo, or if this is a third visit, extract the developed embryo. With men, the procedures are different, on the whole being limited to the extraction of sperm by means of a suction device.

Secret Life is filled with detailed transcripts of Professor Jacobs's interviews with the victims.

The kidnapping, the medical procedures, the forced sex, the production and snatching away of half-human children are horrible and terrifying. But in a profound and very deep way, the worst aspect of it all may be this: what the aliens are doing is unfair. As Professor Jacobs explains:

> Contact between the races is not taking place in a scenario that has been commonly envisioned by scientists and science fiction writers: two independent worlds making careful overtures for equal and mutual benefit. Rather, it is completely one-sided. Instead of equal benefit, we see a disturbing program of apparent exploitation of one species by another. How it began is unknown. How it will end is unknown. But we must face the abduction phenomenon squarely and begin to think rationally about what to do about it.

Professor Jacobs is careful not to overstate his case or to insist that anyone take him and Dr Mack seriously merely because they are professors at prestigious universities. People should believe them, Professor Jacobs explains, because there is no alternative: 'No significant body of thought has come about that presents strong evidence that anything else is happening other than what the abductees have stated.'

Dr Mack agrees, saying '[Dr Jacobs] has made his

case well and has greatly enriched our knowledge of what the abductees have to tell of their experiences. We must go on from here.'

For going on from there, and for bravely telling society unbelievably important news society did not care to hear, John Mack and David Jacobs won the 1993 Ig Nobel Prize in the field of Psychology.

The winners could not, or would not, attend the Ig Nobel Prize Ceremony. Given the gravity of the subject matter, the Ig Nobel Board of Governors arranged for Kevin Steiling, a criminal enforcement official of the Commonwealth of Massachusetts, to address the audience. The audience's nervous laughter came to a halt as Steiling said:

> My name is Kevin Steiling. I am Assistant Attorney General for the Commonwealth of Massachusetts. Kidnapping is a federal offence. It is also a criminal act under the statutes of the Commonwealth of Massachusetts. Last year there were hundreds of kidnappings or attempted kidnappings. None of them involved aliens from other planets. Thank you.

A week after the Ceremony, John Mack's assistant telephoned the Ig Nobel Board of Governors and said 'Dr Mack is delighted that he's won the Ig Nobel Prize. He would like to give a keynote address next year.' She went on to say that Dr Mack's entire staff was excited, and that they planned to buy matching outfits for the occasion.

Over the next twelve months, the Ig Nobel Board of Governors maintained contact with Dr Mack's staff, planning for the occasion. Press releases announced the happy news, and the public expressed excitement and delight at the prospect.

Then, just before the 1994 Ig Nobel Prize Ceremony, Dr Mack's staff informed the Board of Governors that Dr Mack could not, after all attend. At the Ceremony, the master of ceremonies announced:

> We are disappointed and hurt, yes, but above all we are worried. This is an unexplained mystery. Therefore, we are sponsoring a contest to find the true explanation. Please mail your entry, in 25 words or less, to:
>
> 'What Happened to John Mack?'
> c/o The *Annals of Improbable Research*
> Cambridge, Massachusetts 02238
> USA
> Planet Earth
> Solar System
> Milky Way

The Ig Nobel Board of Governors, to its regret, never did learn what happened to John Mack that night.

IGNOBEL PRIZES 2   The Secret Revealed

Professor David Jacobs continued with his research on the alien abduction phenomenon. In the 1992 book with John Mack, he was cautious not to state conclusions, because he was still gathering facts. By 1998, the facts spoke to him decisively, and he published a new book called *The Threat: Revealing the Secret Alien Agenda*. There he bluntly describes what the aliens are doing:

They 'collect human sperm and eggs, genetically alter the fertilized embryo, incubate fetuses in human hosts, and make humans mentally and physically interact with the offspring for proper hybrid development.'

Professor Jacobs explains that the aliens sometimes force abductees to have sex with the hybrids. He is now gathering evidence to prove his suspicion that when enough hybrids exist to assist in a takeover, the aliens will drop the secrecy and make overt contact with us.

John Mack continued to interview people who have been kidnapped by aliens from outer space. In 2000, he published a new book in what promises to be a series. Called *Passport to the Cosmos: Human Transformation and Alien Encounters*, this volume describes the experiences of 100 previously undocumented victims.

Evolution Nyet

KRISHNAS CONGRATULATE KANSAS BOARD OF EDU-CATION
D, K, 16 Oct (VNN) – Monday, 11 October 1999

The Hare Krishnas lauded a landmark decision by the Kansas Board of Education to reject the theory of evolution as a scientific principle in state curricula. Monday's public forum drew citizens, professors, religionists and media alike to a lively debate on this critical issue. Representing the International Society for Krishna Consciousness, Danavir Goswami congratulated the Board by reading a letter from Drutakarma Dasa (a.k.a. Michael A. Cremo) and presenting to them copies of *The Hidden History of the Human Race*, the abridged paperback edition of *Forbidden Archeology*. H.H. Danavir Goswami, director of the Rupa Nuga Vedic College in Kansas City, Missouri, asserted, 'The Board has taken a bold and brilliant step in rejecting Darwin's evolutionary theory as a scientific principle. Darwin bluffed the world with his speculation and double talk, but the Kansas Board of Education has called his bluff.'

— from a 16 October, 1999 report by the Vaishnava News Network

The Official Citation

The Ig Nobel Science Education Prize was awarded to:

The Kansas State Board of Education and the Colorado State Board of Education, for mandating that children should not believe in Darwin's theory of evolution any more than they believe in Newton's theory of gravitation, Faraday's and Maxwell's theory of electromagnetism, or Pasteur's theory that germs cause disease.

IG NOBEL PRIZES 2 Information about evolution is not available from the Kansas State Board of Education, 120 SE 10th Avenue, Topeka, Kansas 66612-1182, USA. Information about evolution is not available from the Colorado State Board of Education, 201 East Colfax Avenue, Denver, Colorado 80203, USA.

The neighbouring Midwest American states of Colorado and Kansas pride themselves on their pioneering heritage. In the late 1990s, the educational authorities of both states tried to pioneer new ways of simplifying science.

It took time for people to accept Isaac Newton's theory of gravity, and, later, for people to accept Einstein's even more-encompassing theory of space, time, and matter. It took time for people to accept Michael Faraday's and James Maxwell's theory of electricity and magnetism. It took time for people to accept Louis Pasteur's theory that germs cause much of the disease that afflicts animals and plants. It took time to accept Gregor Mendel's theory about how characteristics are passed on from parents to their offspring. It took time to accept Charles Darwin's theory about how those characteristics evolve over the course of many generations.

All these theories explain (and are backed up by) a gargantuan, complex pile of evidence. Each theory shows how all of that evidence suddenly fits together and makes sense, once you know the simple principle that explains it.

But they are all just theories. That's all science is – just theories that make sense out of otherwise unrelated facts.

Some people suggest that schools should not expose children to mere theories, that children deserve to hear authoritative facts, not some imperfect explanation of those facts.

In the 1990s, a small group of these people in the United States decided to get some of their members elected to the school boards that decide what is taught and what is not. Voters typically pay little attention to school-board elections, so this would be fairly easy to do. Realizing that others disagreed with them, they adopted a simple strategy: they would downplay or simply not mention their plans until they got elected. In Colorado, and then in Kansas, they did get elected, and then these newly elected, little-noticed public officials quietly voted to start removing theories from the schools – starting with the theory that explains evolution.

For protecting their neighbours' children from theories, the Kansas State Board of Education and the Colorado State Board of Education shared the 1999 Ig Nobel Prize in the field of Science Education.

The winners could not, or would not, attend the Ig Nobel Prize Ceremony. Instead, a concerned citizen from each state came to accept temporary custody of the prizes on behalf of the winners.

From Colorado came young Emily Rosa, who the previous year had come to accept an Ig Nobel Prize on behalf of Dolores Kreiger (the 1998 Ig Nobel Science

Education Prize winner, who was honoured for 'demonstrating the merits of therapeutic touch, a method by which nurses manipulate the energy fields of ailing patients by carefully avoiding physical contact with those patients'). In accepting custody for the Colorado Board of Education she said:

> Well, I'm back. I presented last year's Ig Nobel to Prof. Dolores Krieger who was telling nurses they could heal patients by waving their hands through the air. I'm 12 now, and running into even stranger people. My own school banned the teaching of human evolution. Talking about human evolution is forbidden in my school, as is the origin of all life, and especially the origin of species. It felt like something had blown my whole school into Kansas. Maybe a tornado.

From Kansas came Douglas Ruden, Assistant Professor of Molecular Biosciences at the University of Kansas. In accepting custody for the Kansas Board of Education he said:

> I don't think I'm in Kansas anymore. The Kansas skirmish is the most notorious episode of a long struggle by religious fundamentalists to eliminate the teaching of evolution in public schools. However, this move backfired, because it has caused many teachers in Kansas to actually improve their science teaching. Don't let what happened to Kansas happen to you. As Dan Quayle [the 1998 Ig Nobel Science Education Prize winner, who was honoured for 'demonstrating, better than anyone else, the need for science education'] said, 'A mind is a terrible thing to lose.'

Back in Kansas several days later, Professor Ruden

went to a public meeting of the Kansas Board of Education, where he tried to present the Prize to the actual winners. That meeting was also attended by H.H. Danavir Swami, director of the Rupa Nuga Vedic College in Kansas City, Missouri, who applauded the Board's stand against evolution, and who presented the Board with a congratulatory, freshly-baked cookie. The Board refused the Ig Nobel Prize, but accepted the cookie. The University of Kansas student newspaper reported on the meeting:

> Board members I.B. 'Sonny' Rundell and Val DeFever, who both voted against the new standards, said the board majority deserved the Ig Nobel. 'It deeply saddens me that that is where we are,' DeFever said. 'It embarrasses me to be lumped with them. It makes me furious that their actions have drawn such negative press and thus a negative image of our state.'
>
> Board member Scott Hill, who voted for the standards, said Ruden stooped to a new level in presenting the Ig Nobel prize. 'I'm sure he feels that it helps his position,' he said. 'Misguided, narrow-minded, self-serving bigots generally take the approach to belittle and humiliate their opposition rather than discussing on a factual level.'

Around that same time, the chairman of the Colorado State Board of Education had a gift-wrapped basket of bananas sent to the home of Marc Abrahams, the chairman of the Ig Nobel Board of Governors (and author of this book).

Shortly after that, in Colorado, Emily Rosa's local school board suddenly reversed its policy, and agreed to let evolution be included in biology classes.

The chairman of the Colorado Board of Education sent this token of thanks to the Ig Nobel Board of Governors.

In the next round of elections in both Kansas and Colorado, many – though not all – of the school-board members who voted against evolution were voted out of office.

The Irrelevance of Understanding

Obviously, we now regret having published Sokal's article.
— the editors of the magazine *Social Text*

The Official Citation
The Ig Nobel Literature Prize was awarded to:

The editors of the journal *Social Text*, for eagerly publishing research that they could not understand, that the author said was meaningless, and which claimed that reality does not exist.

Scientists get criticized for using fancy, obscure technical language to describe even the simplest things. But scientists automatically get a lot of respect – the public tends to assume that all scientists are geniuses.

IGNOBEL PRIZES 2 The paper in question was published as 'Transgressing the Boundaries: Toward a Transformative Hermeneutics of Quantum Gravity,' Alan Sokal, *Social Text*, Spring/Summer 1996, pp 217–52. Sokal told the story behind his parody in 'A Physicist Experiments with Cultural Studies,' *Lingua Franca*, May/June 1996, pp 62–4.

Sometimes other academics try to gain that kind of unthinking respect by aping the kind of language they see scientists use. Scientists think this is pretty funny, but they get annoyed at the rare non-scientist who goes not just overboard, but way, way overboard. A few of those overboard professors claim that scientists do nothing more than invent lots of big, technical words, and pretend that those nonsense words mean something. One physics professor got so annoyed at this strange accusation that he decided to do a little something about it.

Professor Alan Sokal of New York University's Physics Department wrote a nonsensical article. Then he sent the article, with a cover letter emphasizing that he was a Big-Time Physics Professor, to a prestigious academic magazine to see if they would publish it.

Sokal's nonsense article is truly a piece of work. Almost nothing in it makes any sense, starting with the title: 'Transgressing the Boundaries: Toward a Trans-formative Hermeneutics of Quantum Gravity'. The paper is a hodgepodge of – yes – big, technical words. It drones on and on, saying nothing, meaning nothing, and spewing lots of jargon and lots of names of famous scientists. There are also footnotes galore, applied with the skill and understanding very young children use the first time they play with mommy's make-up collection.

To Professor Sokal's disgust and glee, the prestigious academic magazine did publish his crazy article.

Then Professor Sokal told the world what had happened, asking 'What's going on here? Could the editors *really* not have realized that my article was written as a parody?'

The answer, he learned later, was 'Yes'.

The news created an uproar. It was reported on the front page of major newspapers around the world. Much amusement was expressed.

But the editors of *Social Text*, who had peacock-proudly published Sokal's nonsense: were not amused. In their next issue they published a reply:

> Whether Sokal's article would have been declared substandard by a physicist peer reviewer is debatable... Sokal's assumption that his parody caught the woozy editors of *Social Text* sleeping on the job is ill-conceived... As for the decision to publish his article, readers can judge for themselves whether we were right or wrong.

Readers were happy to make that judgment.

The whole affair was slightly odd on two counts. First, there *are* scientists who simply make up and use empty jargon, but there aren't many, and they tend to be not very good scientists.

And, second, Professor Sokal's hoax was rather lacking in the courtesy he wanted non-scientists to show.

But the incident did supply much food for many good discussions about what on earth science is all about. And so, for publishing a paper that made sense neither to them nor to its author, the editors of the journal *Social Text* won the 1996 Ig Nobel Prize in the field of Literature.

The winners could not, or would not, attend the Ig Nobel Prize Ceremony, but Alan Sokal, author of the paper, sent them his hearty congratulations, which were read at the Ceremony.

IG NOBEL PRIZES 2   Words Not to Ponder

between the external structure of the physical world and its inner psychological representation *qua* knot theory: this hypothesis has recently been confirmed by Witten's derivation of knot invariants (in particular the Jones polynomial[64]) from

[57]Lacan (1970, 192–193), lecture given in 1966. For an in-depth analysis of Lacan's use of ideas from mathematical topology, see Juranville (1984, chap. VII), Granon-Lafont (1985,1990), Vappereau (1985) and Nasio (1987,1992); a brief summary is given by Leupin (1991). See Hayles (1990, 80) for an intriguing connection between Lacanian topology and chaos theory; unfortunately she does not pursue it. See also Žižek (1991, 38–39, 45–47) for some further homologies between Lacanian theory and contemporary physics. Lacan also made extensive use of concepts from set-theoretic number theory: see e.g. Miller (1977/78) and Ragland-Sullivan (1990).

[58]In bourgeois social psychology, topological ideas had been employed by Kurt Lewin as early as the 1930's, but this work foundered for two reasons: first, because of its individualist ideological preconceptions; and second, because it relied on old-fashioned point-set topology rather than modern differential topology and catastrophe theory. Regarding the second point, see Back (1992).

[59]Althusser (1993, 50): "Il suffit, à cette fin, reconnaître que Lacan confère enfin à la pensée de Freud, les concepts scientifiques qu'elle exige". This famous essay on "Freud and Lacan" was first published in 1964, before Lacan's work had reached its highest level of mathematical rigor. It was reprinted in English translation in 1969 (*New Left Review*).

[60]Miller (1977/78, especially pp. 24–25). This article has become quite influential in film theory: see e.g. Jameson (1982, 27–28) and the references cited there. As Strathausen (1994, 69) indicates, Miller's article is tough going for the reader not well versed in the mathematics of set theory. But it is well worth the effort. For a gentle introduction to set theory, see Bourbaki (1970).

[61]Dean (1993, especially pp. 107–108).

[62]Homology theory is one of the two main branches of the mathematical field called *algebraic topology*. For an excellent introduction to homology theory, see Munkres (1984); or for a more popular account, see Eilenberg and Steenrod (1952). A fully relativistic homology theory is discussed e.g. in Eilenberg and Moore (1965). For a dialectical approach to homology theory and its dual, cohomology theory, see Massey (1978). For a cybernetic approach to homology, see Saludes i Closa (1984).

[63]For the relation of homology to cuts, see Hirsch (1976, 205–208); and for an application to collective movements in quantum field theory, see Caracciolo *et al.* (1993, especially app. A.1).

[64]Jones (1985).

Here is a paragraph from Professor Sokal's article. (The numbers in brackets respond to footnotes, many of which are reproduced in the image above.)

'As Althusser rightly commented, "Lacan finally gives Freud's thinking the scientific concepts that it requires." [59] More recently, Lacan's *topologie du sujet* had been applied fruitfully to cinema criticism [60] and to the psychoanalysis of AIDS. [61] In mathematical terms, Lacan is here pointing out that the first homology group [62] of the sphere is trivial, while those of the other surfaces are profound; and this homology is linked with the connectedness or disconnectedness of the surface after one or more cuts. [63] Furthermore, as Lacan suspected, there is an intimate connection between the external structure of the physical world and its inner psychological representation *qua* knot theory: this hypothesis has recently been confirmed by Witten's derivation of knot invariants (in particular the Jones polynomial [64] from three-dimensional Chern-Simons quantum field theory). [65]'

We suggest that you memorize this paragraph as you might a poem or the lyrics to a song. Recite it the next time you encounter a know-it-all at some party, afterwards pausing momentarily before saying: 'Don't you agree?'

Dianetics

Dianetics is not in any way covered by legislation anywhere, for no law can prevent one man sitting down and telling another man his troubles, and if anyone wants a monopoly on Dianetics, be assured that he wants it for reasons which have to do not with Dianetics but with profit.

— from the book *Dianetics*

The Official Citation

The Ig Nobel Literature Prize was awarded to:

L. Ron Hubbard, ardent author of science fiction and founding father of Scientology, for his crackling Good Book, *Dianetics*, which is highly profitable to mankind or to a portion thereof.

L. Ron Hubbard's classic book.

When a humble science-fiction writer turned his hand to more serious matters, he produced a book that climbed the best-seller charts and stayed there.

The original edition of L. Ron Hubbard's book *Dianetics* began with the statement 'The creation of dianetics is a milestone for Man comparable to his discovery of fire and superior to his invention of the wheel and arch.' The book-buying public apparently agreed – the publishers report that more than eighteen million copies have been sold, and Dianetics has become literally a religion, registered as such for spiritual, legal, and tax purposes.

By many measures, then, *Dianetics* has become a Good Book.

Like anything good, it has its detractors. Right after the book was published, Martin Gumpert reviewed it for the magazine *New Republic*:

> Whatever makes sense in [Hubbard's] 'discoveries' does not belong to him, and his own theory appears to this reviewer as a paranoiac system which would be of interest as part of a case history, but which seems quite dangerous when offered for mass consumption as a therapeutic technique.

Gumpert clearly did not understand the Goodness of what he had read. Perhaps to prevent such confusion, later editions of the book begin with a diplomatic suggestion for the Gumperts of this world:

> In reading this book be very certain you never go past a word you do not fully understand... To help you comprehend the material in this book, words that might be easily misunderstood are defined as footnotes the first time they appear in the text. Each

word so defined has a small number to its right, and the definition appears at the bottom of the page beside the corresponding number.

The footnotes are indeed helpful, and a random sampling hints at the book's wide-ranging subject matter:

PRESENT TIME: the time which is now and becomes the past as rapidly as it is observed. It is a term loosely applied to the environment existing in now.

INTELLIGENCE OFFICER: a military officer responsible for collecting and processing data on hostile forces, weather and terrain.

LOCK: an analytical moment in which the perceptics of the engram are approximated, thus restimulating the engram or bringing it into action, the present time perceptics being erroneously interpreted by the reactive mind to mean that the same condition which produced physical pain once before is now again at hand.

EXODENTISTRY: the extraction of teeth.

Dianetics is a big book – deep, comprehensive, and subtle. It contains far too much wisdom to be understandable at a single reading. Like other great books on which entire religions are based, it has inspired people of great scholarship and spirituality to themselves write books to help new readers interpret the basic text. These commentaries include:

Advanced Procedures and Axioms by L. Ron Hubbard
All About Radiation by L. Ron Hubbard
Assists Processing Handbook by L. Ron Hubbard
The Basic Dianetics Picture Book by L. Ron Hubbard
The Book of Case Remedies by L. Ron Hubbard

The Book of E-meter Drills by L. Ron Hubbard

Child Dianetics by L. Ron Hubbard

Clear Body, Clear Mind by L. Ron Hubbard

The Creation of Human Ability by L. Ron Hubbard

Dianetics 55! by L. Ron Hubbard

Dianetics: the Evolution of a Science by L. Ron Hubbard

The Dynamics of Life by L. Ron Hubbard

E-meter Essentials by L. Ron Hubbard

Group Auditor's Handbook by L. Ron Hubbard

Handbook for Preclears by L. Ron Hubbard

Have You Lived Before This Life? by L. Ron Hubbard

How to Live Through an Executive by L. Ron Hubbard

Introducing the E-meter by L. Ron Hubbard

Introduction to Scientology Ethics by L. Ron Hubbard

Introductory and Demonstration Processes Handbook by
L. Ron Hubbard

Knowingness by L. Ron Hubbard

Notes on the Lectures of L. Ron Hubbard by L. Ron
Hubbard

*The Organization Executive Course and Management
Series* (12 volumes) by L. Ron Hubbard

The Problems at Work by L. Ron Hubbard

Purification: an Illustrated Answer to Drugs by L. Ron
Hubbard

Research and Discovery Series by L. Ron Hubbard

Science of Survival by L. Ron Hubbard

Scientology 0–8 by L. Ron Hubbard

Scientology 8–80 by L. Ron Hubbard

Scientology 8–8008 by L. Ron Hubbard

Scientology: A History of Man by L. Ron Hubbard

Scientology: A New Slant on Life by L. Ron Hubbard

Scientology: The Fundamentals of Thought by L. Ron
Hubbard

Self Analysis by L. Ron Hubbard

The Technical Bulletins of Dianetics and Scientology (18 volumes) by L. Ron Hubbard
Understanding by L. Ron Hubbard
Understanding the E-meter by L. Ron Hubbard

For his contributions to the world of books, L. Ron Hubbard won the 1994 Ig Nobel Prize in the field of Literature.

The winner could not, or would not, attend the Ig Nobel Prize Ceremony. At the Ceremony, MIT astronomer and author Alan Lightman paid tribute to him, saying:

Lafayette Ronald Hubbard was born in 1911. We're not sure when he died. At his peak, L. Ron Hubbard wrote over 100,000 words a month, and I can tell you that's a lot of words. He had a special IBM typewriter with single keys for words like 'the' and 'but', and he had paper that was fed on a roll to avoid wasting time changing sheets. His fiction sold over 23 million copies, and his non-fiction over 27 million. Lastly, Hubbard has written more books posthumously than any writer of our age. He puts living writers to shame.

Following the ceremony, a representative of the Church of Scientology telephoned the Ig Nobel Board of Governors and politely asked, 'Whose idea was it to nominate L. Ron Hubbard for an Ig Nobel Prize?' The Board explained how nominations come in from anyone who wants to send one; how in many cases, including this one, winners are nominated by a large number of people; and that the Board has a tradition (based on an incapacity) of not keeping records. There ensued an interesting, friendly, and pleasant conversation over the next several months.

The Ig was not the only unusual honour for which numerous members of the public have nominated L. Ron Hubbard. In 1988, Random House Publishers asked a board of literary experts to choose the 100 best English-language novels published in the twentieth century, and also asked the reading public to vote. This produced two separate lists. Here are the top ten novels on each of them:

The Board's List

1. *Ulysses* by James Joyce
2. *The Great Gatsby* by F. Scott Fitzgerald
3. *A Portrait of the Artist as a Young Man* by James Joyce
4. *Lolita* by Vladimir Nabokov
5. *Brave New World* by Aldous Huxley
6. *The Sound and the Fury* by William Faulkner
7. *Catch-22* by Joseph Heller
8. *Darkness at Noon* by Arthur Koestler
9. *Sons and Lovers* by D.H. Lawrence
10. *The Grapes of Wrath* by John Steinbeck

The Readers' List

1. *Atlas Shrugged* by Ayn Rand
2. *The Fountainhead* by Ayn Rand
3. *Battlefield Earth* by L. Ron Hubbard
4. *The Lord of the Rings* by J.R.R. Tolkien
5. *To Kill a Mockingbird* by Harper Lee
6. *1984* by George Orwell
7. *Anthem* by Ayn Rand
8. *We the Living* by Ayn Rand
9. *Mission Earth* by L. Ron Hubbard
10. *Fear* by L. Ron Hubbard

The Bible Code

ABSTRACT: It has been noted that when the Book of Genesis is written as two-dimensional arrays, equidistant letter sequences spelling words with related meanings often appear in close proximity. Quantitative tools for measuring this phenomenon are developed. Randomization analysis shows that the effect is significant at the level of 0.00002.

KEY WORDS AND PHRASES: Genesis, equidistant letter sequences, cylindrical representations, statistical analysis.

— from the research report 'Equidistant Letter Sequences in the Book of Genesis'

The Official Citation

The Ig Nobel Literature Prize was awarded to:

Doron Witztum, Eliyahu Rips and Yoav Rosenberg of Israel, and Michael Drosnin of the United States, for their hairsplitting statistical discovery that the Bible contains a secret, hidden code.

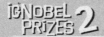

IG NOBEL PRIZES 2

Equidistant Letter Sequences in the Book of Genesis

Doron Witztum, Eliyahu Rips and Yoav Rosenberg

Abstract. It has been noted that when the Book of Genesis is written as two-dimensional arrays, equidistant letter sequences spelling words with related meanings often appear in close proximity. Quantitative tools for measuring this phenomenon are developed. Randomization analysis shows that the effect is significant at the level of 0.00002.

Key words and phrases: Genesis, equidistant letter sequences, cylindrical representations, statistical analysis.

1. INTRODUCTION

The phenomenon discussed in this paper was first discovered several decades ago by Rabbi Weissmandel [7]. He found some interesting patterns in the Hebrew Pentateuch (the Five Books of Moses), consisting of words or phrases expressed in the form of equidistant letter sequences (ELS's)—that is, by se-

elsewhere. Can we now decide between the two possibilities?

Not yet. But suppose now that, aided with the partial dictionary, we can recognise in the text a pair of conceptually related words, like "hammer" and "anvil." We check if there is a tendency of their appearances in the text to be in "close proximity." If

Witztum, Rips and Rosenberg's study was published in *Statistical Science*, volume 9, number 3, 1994, pp 429–38. Drosnin's book *The Bible Code* was published by Simon & Schuster in 1997.

In 1994, three respected scientists announced they had discovered a great and true secret. Someone – they would not specify who – had hidden secret messages in the Bible.

Doron Witztum, Eliyahu Rips and Yoav Rosenberg, hereafter for simplicity's sake called 'W, R & R', are Israeli mathematicians. For years, they devoutly studied the Bible, not for its religious themes or its literary treasures, but to see if it contained secret, coded messages. They found what they were looking for.

To devotees of secret hidden codes in bibles, their method is simply fascinating.

As preparatory steps, W, R & R arranged the letters of the Hebrew version of the Bible as a cylindrical array of row length h. They devised an approximation of the compactness of the configuration. Then they specified the domain of simultaneous minimality for alternate sets of equidistant letter sequences. They then calculated a rough measure of the maximum closeness of the more noteworthy appearances of generic word pairs as equidistant letter sequences in the Book of Genesis.

All this, of course, was mere preparation. The statistical procedure itself is too complex to describe here.

Using their computer as a sort of secret decoder ring, W, R & R decoded the Bible. What they found was, in the words of many who heard about it, unbelievable. W, R & R wrote down a list of 32 names from the *Encyclopedia of Great Men in Israel*. It was all the usual suspects: Rabbi Avraham Ibn-Ezra; Rabbi David of Ganz; Rabbi Heshil of Cracow; the Maharam of Rothenburg; and the others. Then W, R & R had their computer program search through the Bible looking for coded versions of these men's names and coded versions of their birthdays.

They succeeded. They found what they were determined to find.

The three mathematicians wrote a detailed report about their discovery, and got it published in a respected research journal called *Statistical Science*.

This caused a splash, and provoked two kinds of reaction. Some hailed their work as proof that the Bible is a book of unparalleled power and mystery. But most people laughed.

The editor of the journal soon published a special disclaimer: 'Some people seem to think that the publication of the Witztum, Rips and Rosenberg article in *Statistical Science* served as a stamp of scientific approval on the work. This is a great exaggeration... The authors' work did not go far enough to make me seriously think, even for a moment, that their results were anything other than coincidental... The paper was offered to our readers [merely] as a challenging puzzle.'

A new, much larger wave of attention burst in 1979, when Michael Drosnin, a former reporter for the *Wall Street Journal* and the *Washington Post*, came out with his blockbuster book *The Bible Code*. *The Bible Code* described W, R & R's discovery, and went it one better. Drosnin explained that the Bible offers up much more than W, R & R had mentioned. 'The Bible,' he wrote, 'is constructed like a giant crossword puzzle. It is encoded from beginning to end with words that connect to tell a hidden story.'

Drosnin's book opens with a chilling report of a letter he says he had tried to send to Israeli Prime Minister

THE *NEW YORK TIMES* BESTSELLER

THE BIBLE CODE

MICHAEL DROSNIN

Michael Drosnin's best-selling book praises the work of his co-winners Witztum, Rips and Rosenberg, who do not praise the work of Michael Drosnin.

Yitzhak Rabin just days before Rabin was murdered. The letter contained a warning:

> An Israeli mathematician has discovered a hidden code in the Bible that appears to reveal the details of events that took place thousands of years after the Bible was written. The reason I'm telling you this is that the only time your full name – Yitzhak Rabin – is encoded in the Bible, the words 'assassin that will assassinate' cross your name.

Although Drosnin had failed to save Rabin's life, his book was a big hit with the book-buying public.

W, R & R, though, were enraged at this stolen thunder. They held a press conference to say that: 'Mr Drosnin's work employs no scientific methodology... Mr Drosnin's book is based on a false claim. It is impossible to use Torah codes to predict the future.'

Drosnin took the criticism in his stride, and issued a challenge: 'When my critics find a message about the assassination of a prime minister encrypted in *Moby-Dick*, I'll believe them.' Almost immediately, without looking hard at all, Drosnin's critics found messages about the deaths of several prime ministers and of an entire celestial cast of famous people who had made abrupt exits. Among them:

Indira Gandhi
John F. Kennedy
Abraham Lincoln
Reverend Martin Luther King
Yitzhak Rabin
Rene Moawad (President of Lebanon, assassinated
    in 1990)
Engelbert Dollfuss (Chancellor of Austria,
    assassinated in 1934)
Leon Trotsky
Princess Diana and her boyfriend Dodi Fayed and
    their driver Henri Paul

People also found hidden messages in Tolstoy's *War and Peace*, in America's Declaration of Independence, and in pretty much any other long document in which they choose to go secret-code hunting.

Statisticians generally lead a quiet life. Their work, no matter how glamorous, no matter how important, somehow eludes the frenzied media quest to publicize what's hot, what's cool, what's happening. But in this one golden statistical anomaly of a moment, reporters everywhere wanted to interview statisticians, to get their take – to get the lowdown on the Secret Code.

For an oh-so-brief few weeks, statisticians popped up on television news programmes, explaining in methodical detail, with formulae, confidence intervals and other simple tools of their trade, how delightfully easy it is to find all kinds of surprising names and messages in the text of any book, if you don't mind that these names and messages appear *only* when you use some sort of secret decoding scheme. They further pointed out that in any large collection of anything – letters in a book, stars in the sky, scattered refuse from an overturned trash can – you can find, somewhere, almost any pattern you're looking for; there is even a branch of mathematics – called Ramsey Theory – that explains why this is so.

The Bible, though, is a special piece of text, and the secret messages hidden in it are special messages, and Doron Witztum, Eliyahu Rips, Yoav Rosenberg and Michael Drosnin are especially determined, dedicated seekers. Their dedication and determination paid off. The four men shared the 1997 Ig Nobel Prize in the field of Literature.

The winners, all of them, could not, or would not, attend the Ig Nobel Prize Ceremony.

Five years after being honoured with the Ig Nobel Prize, Michael Drosnin published a new book, *The Bible Code II – The Countdown*.

Doron Witztum published his own book, *The Code from Genesis*. He also spent considerable effort responding to responses to his work. Witzum's website, www.torahcodes.co.il, includes some of the detailed papers he wrote about his critics. The lengthy title of one of these papers conveys some of the spirited flavour of Witztum's writing: 'A Response to McKay's Response to My Response to His Response Concerning My Article on...'

430   D. WITZTUM, E. RIPS AND Y. ROSENBERG

FIG. 1.

We call d the *skip*, n the *start* and k the *length* of the ELS. These three parameters uniquely identify the ELS, which is denoted (n, d, k).

Let us write the text as a two-dimensional array—that is, on a single large page—with rows of equal length, except perhaps for the last row. Usually, then, an ELS appears as a set of points on a straight line. The exceptional cases are those where the ELS

FIG. 2.

Witztum, Rips and Rosenberg's rousing research report.

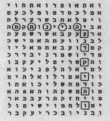

IGNOBEL PRIZES 2   How They Broke the Code

Witztum, Rips and Rosenberg's research report explains how they broke the Bible code, the very existence of which virtually no-one had suspected. Although the entire analysis takes ten pages to describe and involves the use of rather complex mathematical and statistical techniques, their diagrams show at a glance how simple and obvious is the method.

Appendices

Here, appendectally, lie:

The Website

The Annals of Improbable Research

Year-by-Year List of Winners

The Website

About AIR | Subscribe | Submit | What's new? | Events calendar | Contact us

Annals of
IMPROBABLE RESEARCH

HotFlash... **Ig winner Troy Hurtubise**'s grizzly-proof <u>suit of armor is up for sale</u> on Ebay!

Special
Way To Go
Issue

Annals of
IMPROBABLE
RESEARCH

Download a <u>sample issue</u>!

Tox are Airheat **13.1365**
since, January 1, 2004

Our blog: <u>Something New and Improbable</u>, every day M-F

<u>Ig Nobel Prizes</u>
  Our annual awards for achievements that make people Laugh, then Think
<u>AIRchives</u>
    <u>The magazine (AIR)</u> / <u>The newsletter (mini-AIR)</u> / <u>The newspaper column</u>
    ("Improbable Research") / <u>Classics</u> / <u>Press clips</u>
<u>Breaking News & Broken Features</u>
    Our take on current research
<u>Projects & Surveys</u>
    <u>Hair Club</u> / <u>Universal History</u> / <u>Feline Reactions</u>, etc.-- and <u>Improbable</u>
    <u>Research Shows</u>
<u>Unauthorized teaching materials</u>
    Authorized by the <u>Bureaucracy Club</u>
<u>ShareWAIR</u>
    Other improbable websites (submit yours!)
<u>Bookstore</u>
    Improbable books and whatnot

The website.

The Ig Nobel Prizes home page is at www.im-probable.com.

It is part of the *Annals of Improbable Research* website.

There you will find a complete list of the winners and, in many cases, links to their home pages, their original research, press clippings about them, and links to audio and video recordings of many Ig Nobel ceremonies. We also, from time to time, add news of the continuing adventures of past Ig Nobel Prize winners.

FREE NEWSLETTER To keep informed about upcoming Ig

Nobel ceremonies and related events, add yourself to the distribution list for the free monthly newsletter *mini-AIR*. You can do that at the website or, alternatively:

Send a brief e-mail message to this address:

listproc@air.harvard.edu

The body of your message should contain *only* the words

'subscribe mini-AIR' followed by your name. Here are two examples:

subscribe mini-AIR Irene Curie Joliot
subscribe mini-AIR Nicholai Lobachevsky

The magazine.

About the *Annals of Improbable Research*

The *Annals of Improbable Research* (also known as AIR) is a science humour magazine, full of genuine, improbable research culled from more than 10,000 science, medical, and technical, and academic journals, with some deadpan concoctions stirred into the mix of dry humour and juicy pictures. The magazine is printed on old-fashioned paper. There are six issues every year; one of them is devoted to that year's Ig Nobel Prize winners and ceremony.

To see the table of contents of every issue, a selection of classic articles, and an entire free sample issue, go to the Improbable Research website, www.improbable.com.

To subscribe to the magazine, go to the website, or e-mail, phone, or fax:

Annals of Improbable Research
PO Box 380853
Cambridge, MA 02238
USA

Telephone: 617-491-4437
Fax: 617-661-0927
air@improbable.com

Year-by-Year List of Winners

NOTES

1. Additional details about most of the winners – including detailed citations about the published papers and books that were honoured – can be found on the

Improbable Research website, www.improbable.com.
2. In 1991, the very first year that Ig Nobel Prizes were awarded, three additional Prizes were given for apocryphal achievements; those three are not included in the list here. All the other Prizes – including *all* prizes awarded in all subsequent years – were awarded for genuine achievements.

—1991—

ECONOMICS

Michael Milken, titan of Wall Street and father of the junk bond, to whom the world is indebted.

PEACE

Edward Teller, father of the hydrogen bomb and first champion of the Star Wars weapons system, for his lifelong efforts to change the meaning of peace as we know it.

BIOLOGY

Robert Klark Graham, selector of seeds and prophet of propagation, for his pioneering development of the Repository for Germinal Choice, a sperm bank that accepts donations only from Nobellians and Olympians.

CHEMISTRY

Jacques Benveniste, prolific proselytizer and dedicated correspondent of *Nature* for his persistent discovery that water, H_2O, is an intelligent liquid, and for demonstrating to his satisfaction that water is able to remember events long after all trace of those events had vanished.

MEDICINE

Alan Kligerman, deviser of digestive deliverance, vanquisher of vapour, and inventor of Beano, for his pioneering work with anti-gas liquids that prevent bloat, gassiness, discomfort and embarrassment.

EDUCATION

J. Danforth Quayle, consumer of time and occupier of space, for demonstrating, better than anyone else, the need for science education.

LITERATURE

Erich Von Däniken, visionary raconteur and author of *Chariots of the Gods*, for explaining how human civilization was influenced by ancient astronauts from outer space.

—1992—

ECONOMICS

The investors of Lloyd's of London, heirs to 300 years of dull prudent management, for their bold attempt to insure disaster by refusing to pay for their company's losses.

PEACE

Daryl Gates, former Police Chief of the City of Los Angeles, for his uniquely compelling methods of bringing people together.

BIOLOGY

Dr Cecil Jacobson, relentlessly generous sperm donor, and prolific patriarch of sperm banking, for devising a simple, single-handed method of quality control.

ARCHAEOLOGY

Les Eclaireurs de France, the Protestant youth group whose name means 'those who show the way', fresh-scrubbed removers of graffiti, for erasing the ancient paintings from the walls of Mayrières Cave near the French village of Bruniquel.

PHYSICS

David Chorley and Doug Bower, lions of low-energy physics, for their circular contributions to field theory based on the geometrical destruction of English crops.

ART

Presented jointly to Jim Knowlton, modern Renaissance man, for his classic anatomy poster *Penises of the Animal Kingdom*, and to the US National Endowment for the Arts for encouraging Mr Knowlton to extend his work in the form of a pop-up book.

MEDICINE

F. Kanda, E. Yagi, M. Fukuda, K. Nakajima, T. Ohta and O. Nakata of the Shiseido Research Center in Yokohama, for their pioneering research study 'Elucidation of Chemical Compounds Responsible for Foot Malodour,' especially for their conclusion that people who think they have foot odour do, and those who don't, don't.

CHEMISTRY

Ivette Bassa, constructor of colourful colloids, for her role in the crowning achievement of twentieth-century chemistry, the synthesis of bright blue Jell-O.

NUTRITION

The utilizers of Spam, courageous consumers of canned comestibles, for 54 years of undiscriminating digestion.

LITERATURE

Yuri Struchkov, unstoppable author from the Institute of Organoelemental Compounds in Moscow, for the 948 scientific papers he published between the years 1981 and 1990, averaging more than one every 3.9 days.

—1993—

ECONOMICS

Ravi Batra of Southern Methodist University, shrewd economist and best-selling author of *The Great Depression of 1990* ($17.95) and *Surviving the Great Depression of 1990* ($18.95), for selling enough copies of his books to single-handedly prevent worldwide economic collapse.

PEACE

The Pepsi-Cola Company of the Philippines, suppliers of sugary hopes and dreams, for sponsoring a contest to create a millionaire, and then announcing the wrong winning number, thereby inciting and uniting 800,000 riotously expectant winners, and bringing many warring factions together for the first time in their nation's history.

MEDICINE

James F. Nolan, Thomas J. Stillwell and John P. Sands, Jr, medical men of mercy, for their painstaking research report, 'Acute Management of the Zipper-entrapped Penis'.

PHYSICS

Louis Kervran of France, ardent admirer of alchemy, for his conclusion that the calcium in chickens' eggshells is created by a process of cold fusion.

CONSUMER ENGINEERING

Ron Popeil, incessant inventor and perpetual pitchman of late night television, for redefining the industrial revolution with such devices as the Veg-O-Matic, the Pocket Fisherman, Mr Microphone, and the Inside-the-Shell Egg Scrambler.

VISIONARY TECHNOLOGY

Presented jointly to Jay Schiffman of Farmington Hills, Michigan, crack inventor of AutoVision, an image projection device that makes it possible to drive a car and watch television at the same time, and to the Michigan state legislature, for making it legal to do so.

MATHEMATICS

Robert Faid of Greenville, South Carolina, far-sighted and faithful seer of statistics, for calculating the exact odds (710,609,175,188,282,000 to 1) that Mikhail Gorbachev is the Antichrist.

CHEMISTRY

James Campbell and Gaines Campbell of Lookout Mountain, Tennessee, dedicated deliverers of fragrance, for inventing scent strips, the odious method by which perfume is applied to magazine pages.

BIOLOGY

Paul Williams, Jr, of the Oregon State Health Division and Kenneth W. Newell of the Liverpool School of

Tropical Medicine, bold biological detectives, for their pioneering study, 'Salmonella Excretion in Joy-riding Pigs'.

PSYCHOLOGY

John Mack of Harvard Medical School and David Jacobs of Temple University, mental visionaries, for their leaping conclusion that people who believe they were kidnapped by aliens from outer space, probably were – and especially for their conclusion that 'the focus of the abduction is the production of children'.

LITERATURE

E. Topol, R. Califf, F. Van de Werf, P.W. Armstrong and their 972 co-authors, for publishing a medical research paper which has one hundred times as many authors as pages.

—1994—

MEDICINE

This prize is awarded in two parts. First, to Patient X, formerly of the US Marine Corps, valiant victim of a venomous bite from his pet rattlesnake, for his determined use of electroshock therapy: at his own insistence, automobile spark-plug wires were attached to his lip, and the car engine revved to 3000 rpm for five minutes. Second, to Dr Richard C. Dart of the Rocky Mountain Poison Center and Dr Richard A. Gustafson of the University of Arizona Health Sciences Center, for their well-grounded medical report: 'Failure of Electric Shock Treatment for Rattlesnake Envenomation'.

PSYCHOLOGY

Lee Kuan Yew, former Prime Minister of Singapore,

practitioner of the psychology of negative reinforcement, for his thirty-year study of the effects of punishing three million citizens of Singapore whenever they spat, chewed gum, or fed pigeons.

ECONOMICS

Jan Pablo Davila of Chile, tireless trader of financial futures and former employee of the state-owned Codelco Company, for instructing his computer to 'buy' when he meant 'sell', and subsequently attempting to recoup his losses by making increasingly unprofitable trades that ultimately lost 0.5 per cent of Chile's gross national product. Davila's relentless achievement inspired his countrymen to coin a new verb: 'davilar', meaning, 'to botch things up royally'.

PEACE

John Hagelin of Maharishi University and the Institute of Science, Technology and Public Policy, promulgator of peaceful thoughts, for his experimental conclusion that 4,000 trained mediators caused an 18 per cent decrease in violent crime in Washington, DC.

ENTOMOLOGY

Robert A. Lopez of Westport, New York, valiant veterinarian and friend of all creatures great and small, for his series of experiments in obtaining ear mites from cats, inserting them into his own ear, and carefully observing and analysing the results.

PHYSICS

The Japan Meteorological Agency, for its seven-year study of whether earthquakes are caused by catfish wiggling their tails.

MATHEMATICS

The Southern Baptist Church of Alabama, mathematical measurers of morality, for their county-by-county estimate of how many Alabama citizens will go to Hell if they don't repent.

BIOLOGY

W. Brian Sweeney, Brian Krafte-Jacobs, Jeffrey W. Britton and Wayne Hansen, for their breakthrough study, 'The Constipated Serviceman: Prevalence Among Deployed US Troops', and especially for their numerical analysis of bowel-movement frequency.

CHEMISTRY

Texas State Senator Bob Glasgow, wise writer of logical legislation, for sponsoring the 1989 drug-control law which makes it illegal to purchase beakers, flasks, test tubes, or other laboratory glassware without a permit.

LITERATURE

L. Ron Hubbard, ardent author of science fiction and founding father of Scientology, for his crackling Good Book, *Dianetics*, which is highly profitable to mankind or to a portion thereof.

—1995—

PUBLIC HEALTH

Martha Kold Bakkevig of Sintef Unimed in Trondheim, Norway, and Ruth Nielsen of the Technical University of Denmark, for their exhaustive study, 'Impact of Wet Underwear on Thermoregulatory Responses and Thermal Comfort in the Cold'.

DENTISTRY

Robert H. Beaumont, of Shoreview, Minnesota, for his incisive study 'Patient Preference for Waxed or Unwaxed Dental Floss'.

MEDICINE

Marcia E. Buebel, David S. Shannahoff-Khalsa and Michael R. Boyle, for their invigorating study entitled 'The Effects of Unilateral Forced Nostril Breathing on Cognition'.

ECONOMICS

Awarded jointly to Nick Leeson and his superiors at Barings Bank and to Robert Citron of Orange County, California, for using the calculus of derivatives to demonstrate that every financial institution has its limits.

PEACE

The Taiwan National Parliament, for demonstrating that politicians gain more by punching, kicking and gouging each other than by waging war against other nations.

PSYCHOLOGY

Shigeru Watanabe, Junko Sakamoto and Masumi Wakita, of Keio University, for their success in training pigeons to discriminate between the paintings of Picasso and those of Monet.

CHEMISTRY

Bijan Pakzad of Beverly Hills, for creating DNA Cologne and DNA Perfume, neither of which contain deoxy-ribonucleic acid, and both of which come in a triple helix bottle.

PHYSICS

D.M.R. Georget, R. Parker and A.C. Smith, of the Institute of Food Research, Norwich, England, for their rigorous analysis of soggy breakfast cereal, published in the report entitled 'A Study of the Effects of Water Content on the Compaction Behaviour of Breakfast Cereal Flakes'.

NUTRITION

John Martinez of J. Martinez & Company in Atlanta, for Luak Coffee, the world's most expensive coffee, which is made from coffee beans ingested and excreted by the luak (a.k.a. the palm civet), a bobcat-like animal native to Indonesia.

LITERATURE

David B. Busch and James R. Starling, of Madison Wisconsin, for their deeply penetrating research report, 'Rectal Foreign Bodies: Case Reports and a Comprehensive Review of the World's Literature'. The citations include reports of, among other items: seven light bulbs; a knife sharpener; two flashlights; a wire spring; a snuff box; an oil can with potato stopper; eleven different forms of fruits, vegetables and other foodstuffs; a jeweller's saw; a frozen pig's tail; a tin cup; a beer glass; and one patient's remarkable ensemble collection consisting of spectacles, a suitcase key, a tobacco pouch and a magazine.

—1996—

PUBLIC HEALTH

Ellen Kleist of Nuuk, Greenland and Harald Moi of Oslo,

Norway, for their cautionary medical report 'Transmission of Gonorrhoea Through an Inflatable Doll'.

MEDICINE
James Johnston of R.J. Reynolds, Joseph Taddeo of US Tobacco, Andrew Tisch of Lorillard, William Campbell of Philip Morris, Edward A. Horrigan of Liggett Group, Donald S. Johnston of American Tobacco Company and the late Thomas E. Sandefur, Jr, chairman of Brown and Williamson Tobacco Co. for their unshakable discovery, as testified to the US Congress, that nicotine is not addictive.

ECONOMICS
Dr Robert J. Genco of the University of Buffalo, for his discovery that 'financial strain is a risk indicator for destructive periodontal disease'.

PEACE
Jacques Chirac, President of France, for commemorating the fiftieth anniversary of Hiroshima with atomic-bomb tests in the Pacific.

BIODIVERSITY
Chonosuke Okamura of the Okamura Fossil Laboratory in Nagoya, Japan, for discovering the fossils of dinosaurs, horses, dragons, princesses, and more than 1000 other extinct 'mini-species', each of which is less than $\frac{1}{100}$ of an inch in length.

PHYSICS
Robert Matthews of Aston University, England, for his studies of Murphy's Law, and especially for demonstrating that toast often falls on the buttered side.

ART

Don Featherstone of Fitchburg, Massachusetts, for his ornamentally evolutionary invention, the plastic pink flamingo.

CHEMISTRY

George Goble of Purdue University, for his blistering world record time for igniting a barbecue grill – three seconds, using charcoal and liquid oxygen.

BIOLOGY

Anders Barheim and Hogne Sandvik of the University of Bergen, Norway, for their tasty and tasteful report, 'Effect of Ale, Garlic and Soured Cream on the Appetite of Leeches'.

LITERATURE

The editors of the journal *Social Text*, for eagerly publishing research that they could not understand, that the author said was meaningless, and which claimed that reality does not exist.

–1997–

PEACE

Harold Hillman of the University of Surrey, England, for his lovingly rendered and ultimately peaceful report 'The Possible Pain Experienced During Execution by Different Methods'.

MEDICINE

Carl J. Charnetski and Francis X. Brennan, Jr, of Wilkes University, and James F. Harrison of Muzak Ltd in Seattle, Washington, for their discovery that listening to elevator Muzak stimulates immunoglobulin A (IgA) production, and thus may help prevent the common cold.

BIOLOGY

T. Yagyu and his colleagues from the University Hospital of Zurich, Switzerland, from Kansai Medical University in Osaka, Japan, and from Neuroscience Technology Research in Prague, Czech Republic, for measuring people's brain-wave patterns while they chewed different flavours of gum.

ECONOMICS

Akihiro Yokoi of Wiz Company in Chiba, Japan, and Aki Maita of Bandai Company in Tokyo, the father and mother of Tamagotchi, for diverting millions of person-hours of work into the husbandry of virtual pets.

ENTOMOLOGY

Mark Hostetler of the University of Florida, for his scholarly book, *That Gunk on Your Car*, which identifies the insect splats that appear on automobile windows.

ASTRONOMY

Richard Hoagland of New Jersey, for identifying artificial features on the moon and on Mars, including a human face on Mars and ten-mile-high buildings on the far side of the moon.

PHYSICS

John Bockris of Texas A&M University, for his wide-ranging achievements in cold fusion, in the transmutation of base elements into gold, and in the electrochemical incineration of domestic rubbish.

METEOROLOGY

Bernard Vonnegut of the State University of Albany, for his revealing report, 'Chicken Plucking as Measure of Tornado Wind Speed'.

LITERATURE

Doron Witztum, Eliyahu Rips and Yoav Rosenberg of Israel, and Michael Drosnin of the United States, for their hairsplitting statistical discovery that the Bible contains a secret, hidden code.

COMMUNICATIONS

Sanford Wallace, president of Cyber Promotions of Philadelphia – neither rain nor sleet nor dark of night have stayed this self-appointed courier from delivering electronic junk mail to all the world.

–1998–

PEACE

Prime Minister Shri Atal Bihari Vajpayee of India and Prime Minister Nawaz Sharif of Pakistan, for their aggressively peaceful explosions of atomic bombs.

ECONOMICS

Richard Seed of Chicago, for his efforts to stoke up the world economy by cloning himself and other human beings.

STATISTICS

Jerald Bain of Mt Sinai Hospital in Toronto and Kerry Siminoski of the University of Alberta, for their carefully measured report, 'The Relationship Among Height, Penile Length, and Foot Size'.

BIOLOGY

Peter Fong of Gettysburg College, Gettysburg, Pennsylvania, for contributing to the happiness of clams by giving them Prozac.

CHEMISTRY

Jacques Beneveniste of France, for his homeopathic discovery that not only does water have memory, but that the information can be transmitted over telephone lines and the Internet.

SAFETY ENGINEERING

Troy Hurtubise, of North Bay, Ontario, for developing, and personally testing a suit of armour that is impervious to grizzly bears.

MEDICINE

To Patient Y and to his doctors, Caroline Mills, Meirion Llewelyn, David Kelly and Peter Holt, of Royal Gwent Hospital, in Newport, Wales, for the cautionary medical report, 'A Man Who Pricked His Finger and Smelled Putrid for 5 Years'.

SCIENCE EDUCATION

Dolores Krieger, Professor Emerita, New York University, for demonstrating the merits of Therapeutic Touch, a method by which nurses manipulate the energy fields of ailing patients by carefully avoiding physical contact with those patients.

PHYSICS

Deepak Chopra of the Chopra Center for Well Being, La Jolla, California, for his unique interpretation of quantum physics as it applies to life, liberty and the pursuit of economic happiness.

LITERATURE

Dr Mara Sidoli of Washington, DC, for her illuminating report, 'Farting as a Defence Against Unspeakable Dread'.

—1999—

PEACE

Charl Fourie and Michelle Wong of Johannesburg, South Africa, for inventing an automobile burglar alarm consisting of a detection circuit and a flame-thrower.

CHEMISTRY

Takeshi Makino, president of the Safety Detective Agency in Osaka, Japan, for his involvement with S-Check, an infidelity-detection spray that wives can apply to their husbands' underwear.

MANAGED HEALTH CARE

The late George and Charlotte Blonsky of New York City and San Jose, California, for inventing a device (US Patent #3,216,423) to aid women giving birth – the woman is strapped onto a circular table, and the table is then rotated at high speed.

ENVIRONMENTAL PROTECTION

Kyuk-ho Kwon of Kolon Company of Seoul, Korea, for inventing the self-perfuming business suit.

BIOLOGY

Dr Paul Bosland, director of the Chile Pepper Institute, New Mexico State University, Las Cruces, New Mexico, for breeding a spiceless jalapeño chile pepper.

LITERATURE

The British Standards Institution for its six-page specification (BS-6008) of the proper way to make a cup of tea.

SOCIOLOGY

Steven Penfold, of York University in Toronto, for doing his PhD thesis on the sociology of Canadian doughnut shops.

PHYSICS

Dr Len Fisher of Bath, England, and Sydney, Australia, for calculating the optimal way to dunk a biscuit.
    …and…
Professor Jean-Marc Vanden-Broeck of the University of East Anglia, England, and Belgium, for calculating how to make a teapot spout that does not drip.

MEDICINE

Dr Arvid Vatle of Stord, Norway, for carefully collecting, classifying, and contemplating which kinds of containers his patients chose when submitting urine samples.

SCIENCE EDUCATION

The Kansas State Board of Education and the Colorado State Board of Education, for mandating that children should not believe in Darwin's theory of evolution any more than they believe in Newton's theory of gravitation, Faraday's and Maxwell's theory of electromagnetism, or Pasteur's theory that germs cause disease.

—2000—

PSYCHOLOGY

David Dunning of Cornell University and Justin Kreuger of the University of Illinois, for their modest report, 'Unskilled and Unaware of It: How Difficulties in Recognizing One's Own Incompetence Lead to Inflated Self-Assessments'.

PEACE

The British Royal Navy, for ordering its sailors to stop using live cannon shells, and to instead just shout 'Bang!'

CHEMISTRY

Donatella Marazziti, Alessandra Rossi and Giovanni B. Cassano of the University of Pisa, and Hagop S. Akiskal of the University of California (San Diego), for their discovery that, biochemically, romantic love may be indistinguishable from having severe obsessive-compulsive disorder.

ECONOMICS

The Reverend Sun Myung Moon, for bringing efficiency and steady growth to the mass-marriage industry, with, according to his reports, a 36-couple wedding in 1960, a 430-couple wedding in 1968, an 1800-couple wedding in 1975, a 6000-couple wedding in 1982, a 30,000-couple wedding in 1992, a 360,000-couple wedding in 1995, and a 36,000,000-couple wedding in 1997.

MEDICINE

Willibrord Weijmar Schultz, Pek van Andel and Eduard Mooyaart of Groningen, the Netherlands, and Ida

Sabelis of Amsterdam, for their illuminating report, 'Magnetic Resonance Imaging of Male and Female Genitals During Coitus and Female Sexual Arousal'.

PUBLIC HEALTH

Jonathan Wyatt, Gordon McNaughton and William Tullet of Glasgow, Scotland for their alarming report, 'The Collapse of Toilets in Glasgow'.

PHYSICS

Andre Geim of the University of Nijmegen, the Netherlands, and Sir Michael Berry of Bristol University (UK), for using magnets to levitate a frog.

COMPUTER SCIENCE

Chris Niswander of Tucson, Arizona, for inventing Paw-Sense, software that detects when a cat is walking across your computer keyboard.

BIOLOGY

Richard Wassersug of Dalhousie University, for his first-hand report, 'On the Comparative Palatability of Some Dry-Season Tadpoles from Costa Rica'.

LITERATURE

Jasmuheen (formerly known as Ellen Greve) of Australia, first lady of Breatharianism, for her book *Living on Light*, which explains that although some people do eat food, they don't ever really need to.

—2001—

PUBLIC HEALTH

Chittaranjan Andrade and B.S. Srihari of the National Institute of Mental Health and Neurosciences, Banga-lore, India, for their probing medical discovery that nose picking is a common activity among adolescents.

PSYCHOLOGY
Lawrence W. Sherman of Miami University, Ohio, for his influential research report 'An Ecological Study of Glee in Small Groups of Preschool Children'.

ECONOMICS
Joel Slemrod, of the University of Michigan Business School, and Wojciech Kopczuk, of the University of British Columbia, for their conclusion that people find a way to postpone their deaths if that would qualify them for a lower rate on the inheritance tax.

PEACE
Viliumas Malinauskus of Grutas, Lithuania, for creating the amusement park known as Stalin World.

MEDICINE
Peter Barss of McGill University, for his impactful medical report 'Injuries Due to Falling Coconuts'.

PHYSICS
David Schmidt of the University of Massachusetts, for his partial solution to the question of why shower curtains billow inwards.

TECHNOLOGY
Awarded jointly to John Keogh of Hawthorn, Victoria, Australia, for patenting the wheel in the year 2001, and to the Australian Patent Office for granting him Innovation Patent #2001100012.

ASTROPHYSICS
Dr Jack and Rexella Van Impe of Jack Van Impe Min-

istries, Rochester Hills, Michigan, for their discovery that black holes fulfil all the technical requirements to be the location of Hell.

BIOLOGY

Buck Weimer of Pueblo, Colorado, for inventing Under-Ease, airtight underwear with a replaceable charcoal filter that removes bad-smelling gases before they escape.

LITERATURE

John Richards of Boston, England, founder of The Apostrophe Protection Society, for his efforts to protect, promote, and defend the differences between plural and possessive.

–2002–

BIOLOGY

Norma E. Bubier, Charles G.M. Paxton, Phil Bowers and D. Charles Deeming of the United Kingdom, for their report 'Courtship Behaviour of Ostriches Towards Humans Under Farming Conditions in Britain'.

PHYSICS

Arnd Leike of the University of Munich, for demonstrating that beer froth obeys the mathematical Law of Exponential Decay.

INTERDISCIPLINARY RESEARCH

Karl Kruszelnicki of the University of Sydney, for performing a comprehensive survey of human belly-button lint – who gets it, when, what colour and how much.

CHEMISTRY

Theodore Gray of Wolfram Research, in Champaign, Illinois, for gathering many elements of the periodic table, and assembling them into the form of a four-legged periodic-table table.

MATHEMATICS

K.P. Sreekumar and the late G. Nirmalan of Kerala Agricultural University, India, for their analytical report 'Estimation of the Total Surface Area in Indian Elephants'.

LITERATURE

Vicki L. Silvers of the University of Nevada-Reno and David S. Kreiner of Central Missouri State University, for their colourful report 'The Effects of Pre-existing Inappropriate Highlighting on Reading Comprehension'.

PEACE

Keita Sato, President of Takara Co., Dr Matsumi Suzuki, President of Japan Acoustic Lab, and Dr Norio Kogure, Executive Director, Kogure Veterinary Hospital, for promoting peace and harmony between the species by inventing Bow-Lingual, a computer-based automatic dog-to-human language translation device.

HYGIENE

Eduardo Segura, of Lavakan de Aste, in Tarragona, Spain, for inventing a washing machine for cats and dogs.

ECONOMICS

The executives, corporate directors and auditors of Enron, Lernout & Hauspie (Belgium), Adelphia, Bank of Commerce and Credit International (Pakistan),

Cendant, CMS Energy, Duke Energy, Dynegy, Gazprom (Russia), Global Crossing, HIH Insurance (Australia), Informix, Kmart, Maxwell Communications (UK), McKessonHBOC, Merrill Lynch, Merck, Peregrine Systems, Qwest Communications, Reliant Resources, Rent-Way, Rite Aid, Sunbeam, Tyco, Waste Management, WorldCom, Xerox, and Arthur Andersen, for adapting the mathematical concept of imaginary numbers for use in the business world. (Note: all companies are US-based unless otherwise noted.)

MEDICINE

Chris McManus of University College London, for his excruciatingly balanced report, 'Scrotal Asymmetry in Man and in Ancient Sculpture'.

—2003—

ENGINEERING

The late John Paul Stapp, the late Edward A. Murphy, Jr, and George Nichols, for jointly giving birth in 1949 to Murphy's Law, the basic engineering principle that 'If there are two or more ways to do something, and one of those ways can result in a catastrophe, someone will do it' (or, in other words: 'If anything can go wrong, it will').

PHYSICS

Jack Harvey, John Culvenor, Warren Payne, Steve Cowley, Michael Lawrance, David Stuart and Robyn Williams of Australia, for their irresistible report 'An Analysis of the Forces Required to Drag Sheep over Various Surfaces'.

MEDICINE

Eleanor Maguire, David Gadian, Ingrid Johnsrude, Catriona Good, John Ashburner, Richard Frackowiak and Christopher Frith of University College London, for presenting evidence that the brains of London taxi drivers are more highly developed than those of their fellow citizens.

PSYCHOLOGY

Gian Vittorio Caprara and Claudio Barbaranelli of the University of Rome, and Philip Zimbardo of Stanford University, for their discerning report 'Politicians' Uniquely Simple Personalities'.

CHEMISTRY

Yukio Hirose of Kanazawa University, for his chemical investigation of a bronze statue, in the city of Kanazawa, that fails to attract pigeons.

LITERATURE

John Trinkaus, of the Zicklin School of Business, New York City, for meticulously collecting data and publishing more than 80 detailed academic reports about things that annoyed him, such as: what percentage of young people wear baseball caps with the peak facing to the rear rather than to the front; what percentage of pedestrians wear sport shoes that are white rather than some other colour; what percentage of swimmers swim laps in the shallow end of a pool rather than the deep end; what percentage of automobile drivers almost, but not completely, come to a stop at one particular stop sign; what percentage of commuters carry attaché cases; what percentage of shoppers exceed the number of items permitted in a supermarket's express checkout lane; and what percentage of students dislike the taste of Brussels sprouts.

ECONOMICS

Karl Schwärzler and the nation of Liechtenstein, for making it possible to rent the entire country for corporate conventions, weddings, bar mitzvahs, and other gatherings.

INTERDISCIPLINARY RESEARCH

Stefano Ghirlanda, Liselotte Jansson and Magnus Enquist of Stockholm University, for their inevitable report 'Chickens Prefer Beautiful Humans'.

PEACE

Lal Bihari, of Uttar Pradesh, India, for a triple accomplishment: First, for leading an active life even though he has been declared legally dead; Second, for waging a lively posthumous campaign against bureaucratic inertia and greedy relatives; and Third, for creating the Association of Dead People.

BIOLOGY

C.W. Moeliker, of Natuurmuseum Rotterdam, the Netherlands, for documenting the first scientifically recorded case of homosexual necrophilia in the mallard duck.